T0213529

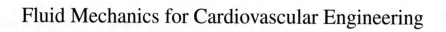

Fluid Mechanics for Cardiovascular Engineering

Gianni Pedrizzetti

Fluid Mechanics for Cardiovascular Engineering

A Primer

 Springer

Gianni Pedrizzetti
Department of Engineering and Architecture (DIA)
University of Trieste
Trieste, Italy

ISBN 978-3-030-85945-9 ISBN 978-3-030-85943-5 (eBook)
https://doi.org/10.1007/978-3-030-85943-5

Cover image: Steady streaming flow in a healthy left ventricle, streamlines show the blood flow pattern
during one heartbeat; brighter colors correspond to a higher velocity. Data obtained by numerical solution
of the equations governing blood motion inside a cardiac geometry extracted from clinical images.

This Springer imprint is published by the registered company Springer Nature Switzerland AG
The registered company address is: Gewerbestrasse 11, 6330 Cham, Switzerland

To my wife Ilaria,

the better half of me,

and our three precious children

Preface

Fluid mechanics is gaining increasing relevance in cardiovascular sciences since it is recognized that most cardiovascular dysfunctions are associated with the alteration of blood flow in the heart chambers, across cardiac valves, and inside major vessels.

The study of fluid mechanics in such districts of pathophysiological interest necessities an encounter between competencies that range from a rigorous physical-mathematical background on fluid dynamics to the careful understanding of the medical field of application, clinical cardiology. This is an exemplary situation where it is appropriate to talk of a multidisciplinary approach, because mechanics and cardiology are very distant fields, both in their technical content and in their cultural roots, and there are no well-defined interdisciplinary fields lying somewhere between them. Several disciplines must be involved and such a purely multidisciplinary field faces the challenge of building a fertile collaborative ground maintaining the expertise of individual disciplines and a common language for efficient communication. On the one hand, expert scientists have to make an effort to understand in depth the requirements coming from the clinical or physiological field, without departing from their scientific rigorous roots. On the other hand, clinical researchers must try to understand the potential offered by scientific disciplines, yet maintaining a deep involvement in the clinical applications. Working synergistically, they can formulate the questions that can be solved and that are important for improving clinical cardiology. I worked in this multidisciplinary territory for more than 25 years now, coming from my studies in theoretical fluid mechanics, and had the luck to be guided in the realm of cardiology by my dear friend, Dr. Giovanni Tonti, a top-rated cardiologist with a sharp mind eagerly open to basic science and technological innovation. Thanks to our reciprocal contamination, we could work efficiently in this multi-disciplinary field; a field that is acquiring growing relevance in academic and industrial systems. This book aims to describe this multi-disciplinary field as experienced by the engineering side.

This is a book about fluid dynamics. It is a textbook based on an extensive revision of the lecture notes of my course held at the University of Trieste to the fourth-year students in clinical, biomedical, and mechanical engineering. It is an introductory course on fluid mechanics that runs along a common thread that eventually leads to cardiovascular applications. As such, it starts with rigorous physical foundations of fluid dynamics although read with a perspective to their applications to the circulation. It then contains classic and specific results that are pertinent for the application of fluid dynamics to cardiovascular physiology. After that, the text introduces the fundamental elements of vortex dynamic that supply the reader with physics-based interpretative models for those phenomena effectively encountered in clinical cardiology, which are analyzed in the final, clinical-oriented chapters. All arguments are presented at a fundamental level and include original insights. Therefore, potential readers are research scientists (physicists, engineers) working in cardiovascular fluid dynamics, industry engineers working on biomedical/cardiovascular technology, and students in the various branches of bioengineering including biomedical and clinical engineering.

Along this line, I avoided to scare the reader with encyclopedic material on specific aspects—already available in some textbooks of specific subjects—rather ensuring the information necessary to operate with scientific rigor in the clinical applications. By my personal attitude, the book is written using an essential style, aimed to provide a thread along the material treated, from theoretical background to applications in clinical cardiology. However, the topic is so wide that coverage of arguments is certainly incomplete, many important topics have been overlooked and the clinical arguments are intentionally introduced at a superficial level just to provide a starting point to unexperienced, non-medical readers. The principal objective is one of creating an initial fertile ground to support the growth in this important subject that merges theoretical rigor with the uncertainty arising from the complexity of clinical cardiology.

The book is subdivided into 4 conceptually separate parts in sequence.

The Part I contains the preparatory material. Fundamental concepts include the definition of fluid and solids as models to describe physical phenomena and dimensional congruence as a powerful tool that is often overlooked. The background includes the statics of fluid and the methods to properly describe fluid motion. The latter includes an intuitive description of mathematical operators and underlines the value of the Gauss theorem.

The Part II is dedicated to the rigorous formulation of the conservation laws (mass, momentum, energy) that will be used in the rest of the book. The laws are formulated in integral form, in their one-dimensional formulation for application to vessels and in the general differential form. Results are accompanied by some examples that are pertinent to the cardiovascular system.

The Part III reports the main, somehow classic results of fluid dynamics that are pertinent to the circulatory system, including propagation of elastic waves and branched and curved geometry. These analytical results are necessarily limited to simple flow conditions where blood motion is mostly unidirectional. They are accompanied by examples or brief discussions regarding their physiological meaning.

Part IV provides the material for understanding effective physiological flows that present boundary layer separation and vortex formation. These represent the main situations of pathophysiological significance in the heart and large blood vessels. This part contains a preparatory chapter with a theoretical background on the dynamics of vorticity. This provides interpretative models to discuss the complexity of blood flow in the application discussed later: large vessels, cardiac chambers, and across cardiac valves. This part is accompanied by some background on clinical cardiology and by discussions on the significance of fluid dynamics in the clinical context.

The material contained in this textbook is certainly incomplete, the mathematics is kept at a basic level often favoring intuition over rigor, and the clinical parts are oversimplified. Moreover, the written text can contain numerous inaccuracies and stylistic poorness due to the writing by a non-native speaker. I tried my best; for sure, all these shortcomings are imputable to me only, either for my imprudent choices or my limitations.

Prato, Italy   Gianni Pedrizzetti
7 September 2021

Contents

Part I
Introductory Elements

Chapter 1
Basic Concepts

Abstract This chapter provides an overview of the fundamental concepts required to study fluid mechanics in the cardiovascular system. It starts with the definition of what is intended with mechanics and with the introduction of the concept of the continuum. Fluid and solids are discussed as models of continuous material and their differences are described in terms of their parametric and dynamical properties. An outline of the main biological fluid domains is presented with a description of the elements of the cardiovascular systems. In conclusion, dimensional analysis is introduced, as a foundation for the representation of physical properties and a powerful constraint underlying all physical laws.

1.1 Mechanics and Continuum

"Mechanics" can be described in general as that field of science that is dedicated to the study of the laws of motion of physical bodies or, equivalently, to the equilibrium of the momentum of their constitutive elements. Mechanics can be divided into three main subjects:

1. *Statics* that deals with the equilibrium of all forces applied to a physical system in absence of any motion.
2. *Kinematics* that deals with the methods to describe the motion irrespective of the applied forces that create such motion.
3. *Dynamics*, the most important and comprehensive part regarding the relationship between forces and motion.

The term "dynamics" derives from the ancient Greek (δυναμικός) and was renewed in the French word *dynamique* by Leibnitz (1646–1716), where it got the meaning of "pertaining with forces producing motion". The concept of mechanics was addressed in mathematical terms by Newton who demonstrated that forces are the entities that change the motion, because they produce acceleration. Newton laws of classical mechanics were carefully developed for individual point particles of finite mass; they were then extended to rigid bodies and to deformable material. Here we will have to revise these classical laws of mechanics for their applications to fluid elements.

© The Author(s), under exclusive license to Springer Nature Switzerland AG 2022

G. Pedrizzetti, *Fluid Mechanics for Cardiovascular Engineering*,

https://doi.org/10.1007/978-3-030-85943-5_1

This book is about classical mechanics; it will ignore modern developments of quantum mechanics whose corrections are largely negligible for objects of size much larger than an individual sub-atomic constituent of matters. It also neglects relativity correction because fluids within the body move with velocities well below the speed of light. The focus is maintained to pure mechanics and discussions regarding thermic and chemical phenomena are intentionally avoided, with the exception of a few mentions that are reported where appropriate.

The course is specifically about biological fluids, which are mainly air, water, and blood. However, it is important to remark from the beginning that the concept of "fluid" is a "model" used to describe certain phenomena encountered in the real world. Furthermore, fluids and solids represent the main classes of the wider model commonly described as a "continuum". No material is really a continuum, it is made of individual molecules that are made of atoms, which are made of sub-atomic particles; however, the model of continuum is used to describe macroscopic phenomena whose modification occurs on scales that are incommensurably larger than those of such individual constituents.

Air and water have molecules whose size is of the order of nanometers (1 nm = 10^{-9} m); for them, the scheme of continuum is appropriate when studying macroscopic phenomena whose size is much larger than that. For water and air, macroscopic scales can be, in absolute terms, as small as little fractions of a millimeter.

Blood is different; blood is a particulate fluid mixture composed by a percentage of about 50% by plasma (that can be considered analogous to water to a good approximation) and another about 50% of red blood cells (this percentage is called hematocrit), plus minor percentages of white cells and other constituents. Red blood cells, transport oxygen in the whole body and they are much larger than water molecules. They have a discoidal shape of radius of about 8 μm (1 μm = 10^{-6} m), thicker around the circumference, with a thickness of about 2 μm, and a thin membrane at the center. Its shape may roughly be described as a donut, whose hole is covered by a membrane that extends from the surface of the inside ring to the center (see Fig. 1.1). Thus, the volume of a red blood cell is approximately 10^{-7} mm^3 and, if blood cells

Fig. 1.1 Red blood cell

cover 50% of blood volume, there are about 5×10^6 red blood cells in one mm^3 of blood.

Based on these figures, blood motion should be described with a corpuscular or a continuous model depending on the size of the domain under analysis, for example, the diameter of the vessel. Large blood vessels have a diameter ranging from a few centimeters to several millimeters, here the continuum model is appropriate. Smaller vessels have a size that can contain some tens of red blood cells across, here the corpuscular nature of blood presents a certain influence. At the smaller end, the diameter of capillaries is less than 10 μm; here red blood cells flow one after the other in a row, even squeezing to be able to pass through, and the corpuscular nature of blood takes a fundamental relevance.

However, the physiological sites of greater clinical interest, where the mechanical phenomena present a direct physiological counterpart, are the heart chambers and the large vessels, like aorta and carotid for example. In the heart and large vessels, blood dynamics can be confidently modeled as that of continuous fluid. However, even in such large vessels, a few specific phenomena may still be influenced by blood's corpuscular nature and, when such phenomena present a physiological significant, they should be properly accounted.

For most practical applications, the simplified representation of blood as a continuous material allows employing a rich theoretical background of continuous mechanics and differential mathematics that represents the basic tools of most achievements in fluid dynamics.

The continuous model is appropriate for describing the large-scale phenomena of motion, when changes in the fluid motion occur over distances that are orders of magnitudes larger than individual constituents. A continuous mean is characterized by either its global properties or its local properties. Examples of global, or integral, properties are the volume V or the mass M of the portion of material under analysis. The density ρ of a volume of fluid is given by the ratio M/V, mass per unit volume. However, the density of a volume of fluid represents an average value over that volume of a local property, because density is a property that can take different values at different positions inside the volume. The density can be defined locally at every point as

$$\rho = \lim_{V \to 0} \frac{M}{V} = \frac{dM}{dV}. \tag{1.1}$$

The second equality in (1.1) used the differential form of a ratio between infinitesimal quantities which implicitly assumes the limit $dV \to 0$. Here, it should be remarked once again that in the continuous model the infinitesimal volume is still much larger than the individual constituents of the material. Needless to remind that global properties can be evaluated by integration of local ones, like the mass of a volume is given by the integral of the density over such volume

$$M = \int_V \rho \, dV. \tag{1.2}$$

Local properties provide the most comprehensive description of the continuum; they are also called "fields". Fields are mathematical quantities that take different values at different points in space and can also vary in time, like temperature $T(x, t)$ or pressure $p(x, t)$, where t is the time coordinate and x is the space coordinate vector; similarly, velocity $v(x, t)$ is a vector field.

The physical laws that govern the mechanics of a continuum are the "conservation of mass" and the "conservation of momentum", including angular momentum, which are the expressions of the Newton law. Given the limited changes in temperature inside the circulatory system, we simplify the whole matter by neglecting thermodynamics phenomena. We also assume that the material does not undergo transformations (of state, chemical, or else) and maintains the same properties as time progresses. Under these simplified conditions, the only form of energy coming into play is mechanical energy in its manifestations of kinetic and potential energy. Other forms of energy like those associated with heat transport or chemical reactions are thus neglected; this means that any non-mechanical property, like temperature or concentration of a solute, does not influence the motion actively and it is transported passively with the fluid. In this purely mechanical scenario, where the only form of energy is mechanical energy, the conservation of momentum can be recast to express the "conservation of energy" that is not an additional conservation law.

Conservation laws must be integrated with an "equation of state" that characterizes the behavior of the specific continuous material under analysis. The equation of state is the law that relates volume V, pressure p, and temperature T. A well-known example is the law $pV = RT$ of ideal gas. In a continuum, it is preferable to express the equation of state entirely in terms of local variable, or fields, and rewrite it as a relation between density, pressure, and temperature

$$\rho = f(p, T). \tag{1.3}$$

The role of temperature in (1.3) is typically associated with the decrease of density in regions where temperature increases. This effect, besides local sources of thermal energy, is important in large regions like the atmosphere or the oceans where the large differences in temperature can give rise to stratification of density with the quote and are responsible for buoyancy effects. Differently, inside the human body temperature can be confidently considered constant or non-influent such that (1.3) reduces to its isothermal version

$$\rho = f(p), \tag{1.4}$$

which expresses the intuitive phenomenon that density increases when pressure increases and vice versa.

Let us quantify this point a little more carefully: consider a generic volume V of material (for example, a cylinder), that is in equilibrium with the external pressure, and apply a pressure increment dp on the surrounding surface of such volume (for example, pushing a piston in the cylinder). The volume will be compressed and experience a (negative) change of volume, dV, that increases with increasing pressure dp. The compressibility of a material is thus measured by the ratio between the increase of pressure and the corresponding decrease of volume, which must be expressed relative to its initial volume V because the same pressure acts on every infinitesimal element. This ratio is known as the modulus of cubic compressibility (or bulk modulus). It is expressed as

$$\varepsilon = -\frac{dp}{dV/V} = \rho\frac{dp}{d\rho}, \tag{1.5}$$

which is larger for stiffer materials, where a large increase of pressure is required to have a small decrease of volume (the negative sign is included to have a positive value of ε because changes of pressure and volume have always opposite signs). The second equality in (1.5) used the relation between volume and density, $V = \frac{M}{\rho}$, with the mass M being a constant, to transform volume variations into density ones. The previous equation can be rearranged to express the relative change of density

$$\frac{d\rho}{\rho} = \frac{dp}{\varepsilon} = \frac{dp}{\rho c^2}, \tag{1.6}$$

where the last equality used the definition of velocity of propagation of sound (Kundu et al., 2012), $c = \sqrt{dp/d\rho}$, in the material that gives $\varepsilon = \rho c^2$.

In liquid materials, the modulus ε is commonly large because a small reduction of volume can be achieved only in the presence of extremely large increases of pressure that are typically not physiological. Change in pressure can also develop by changes of fluid flow, for example, when a fast-moving fluid impacts onto a still surface, thus decelerating its velocity from a value v to zero, the pressure variation dp is proportional to ρv^2 (details will be given in Chap. 6). Therefore, using (1.6), the relative change in density $d\rho/\rho$ turns out to be proportional to the ratio between the squares of fluid velocity and velocity of sound, $(v/c)^2$. Physiological velocities are much lower than the velocity of sound (that in water is about 1500 m/s and in dry air about 345 m/s); the condition that $v \ll c$, amplified for square velocities, implies that $dp \ll \rho$. In summary, the variation of density induced by physiological values of either pressure or blood velocity is largely negligible; therefore, we can focus the attention on the limiting case of "incompressible material", where density is assumed as a constant property (or $\varepsilon \to \infty$). Therefore, the equation of state to our purpose takes the simple form

$$\rho = constant, \tag{1.7}$$

where the density takes values about 10^3 kg/m^3 in water, 1.05×10^3 kg/m^3 for blood, and about 1.2 kg/m^3 in air at 20 °C and atmospheric pressure.

1.2 Fluids and Solids

The discussion brought forward so far applies to a generic continuum and does not make explicit use of the effective nature of the material, which can be either a solid or a fluid. Before moving ahead, it is time to elucidate the difference between solids and fluid so that we can then focus on the latter with no ambiguity.

A solid material, such as biological hard tissue like a bone or a soft tissue like a muscle, presents its own shape due to a natural geometric organization of the constituting elements, which are due to the bonds between their atomic or molecular components. When the relative position of these elements is altered through a small amount of differential displacement, internal stresses develop in an effort to restore the elements to their original, stress-free state. For example, with reference to Fig. 1.2, a rod of elastic material of length L stretches under the action of an external force F because the deformation of elements generates a field of internal forces that counterbalance the force F. When the force ceases, the deformation goes back to zero and the rod returns to its original length. This distinctive property of solids, where internal stresses develop in response to a deformation, is generally called "elasticity". In its simplest form, elastic energy is a form of potential energy, which is stored in the deformed structure composing the material (like if neighboring elements were connected by small springs) and is returned when the deformation goes back to zero. Indeed, an elastic deformation is normally completely reversible. In the simple one-dimensional case of Fig. 1.2, the elastic deformation, or strain, $s = \Delta L / L$, is related to the amount of stress, $\tau = F/A$, proportional to the force and inversely proportional to the area of the cross section. In general, solid materials are characterized by a stress–strain relationship as shown in Fig. 1.3, which represents the "constitutive law" characterizing the elastic behavior of a solid material. For small enough deformation, the stress–strain relationship can be considered as linear and, for one-dimensional deformation, it is written $\tau = Es$, where the proportionality coefficients E is the Young modulus. Most biological tissues, however, are subjected to relatively large deformation and the linear behavior is only an approximation. The stress–strain relationship in biological tissues, sketched in Fig. 1.3, features a sharp increase of stiffness when the material is subjected to increasingly large deformations; such behavior represents a protective property that limits the entity of deformation in the event of extreme overloads.

Fig. 1.2 Elasticity in solids: material deforms under the action of a force

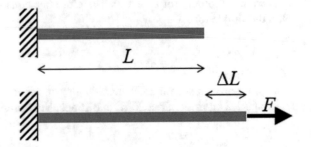

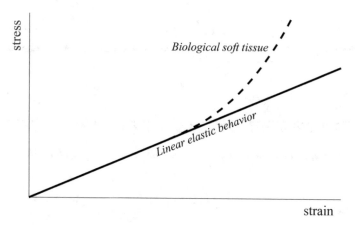

Fig. 1.3 Elasticity in solids: stress–strain relationship

Fluids are different and do not present an elastic behavior. The most distinctive property of fluids, which include liquids and gases, is that a fluid has not a preferred shape. A fluid offers no resistance to taking the shape of its container irrespective of any geometry it had previously. The individual elements (e.g., atoms, molecules) constituting a fluid have not preferred relative positions; thus, they may be organized in infinitely many stress-free states. Differently from solids, fluids do not develop stresses for a relative displacement of their constituting elements; instead, fluids develop an internal resistance during their relative motion. Indeed, the distinctive property of fluids is the development of internal stresses in response to a "rate of deformation", to a differential velocity between nearby elements. This property of fluids takes the name of "viscosity". A fluid thus experiences a viscous resistance during the motion caused by the sliding of the individual fluid elements one on the other. Viscous stresses represent a frictional phenomenon that appears during motion, when the motion ceases also the viscous stress ceases and there is no mechanism taking the system back to its original configuration as it happened in solids. The mechanical energy used to deform the fluid elements has not been stored anywhere, it is dissipated by internal viscous friction and irreversibly transformed into heat and dispersed.

In analogy to what was previously shown for elasticity in solids, fluids are characterized by a relationship between stress and rate-of-strain. Consider a simple experiment of a thin layer of fluid between two walls (infinitely extended to avoid end-effects), the lower wall being fixed and the upper wall sliding with constant velocity U, as sketched in Fig. 1.4. The upper wall is maintained at constant velocity under the action of a shear action τ, given by the force per unit area. Such shear force is proportional to U (it increases when velocity increases) and inversely proportional to the thickness d (it increases when thickness decreases); eventually it can be shown to depend on the ratio U/d. When the thickness is small enough, such ratio approximately corresponds to the velocity derivative and, with reference to Fig. 1.4, one can write

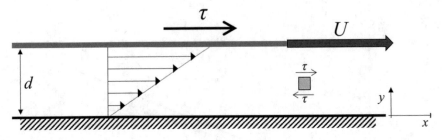

Fig. 1.4 Viscosity in fluids: shear frictions between fluid elements sliding with different velocity

$$\tau = f\left(\frac{U}{d}\right) = f\left(\frac{dv_x}{dy}\right).$$ (1.8)

This relationship between shear stress and shear rate is the "constitutive law" that characterizes the viscous fluid properties as shown in Fig. 1.5 for a few examples. Fluids that follow a simple linear relationship are called "Newtonian fluids" for which (1.8) takes the simple form

$$\tau = \mu \frac{dv_x}{dy},$$ (1.9)

and the proportionality coefficient μ is the "dynamic viscosity", a property that characterizes the amount of viscous resistance to fluid motion. Fortunately, most common fluids like water and air closely follow the behavior of a Newtonian fluid; dynamic viscosity in water takes value about 10^{-3} kg/m·s at 20° that decreases to about 0.7 10^{-3} kg/m·s at 37°, and 1.8 × 10^{-5} kg/m·s in air at 20°.

Blood is more similar to a shear-thinning fluid because its corpuscular nature influences the value of viscosity, which cannot be assumed circumstantially constant.

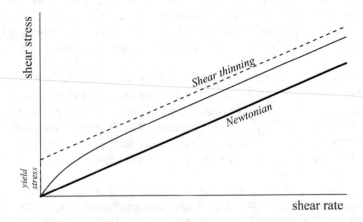

Fig. 1.5 Viscosity in fluids: shear stress depends on shear rate

In fact, being blood a mixture of corpuscular elements into an aqueous solution, its apparent viscosity is not an intrinsic material property and the value depends on the type of motion that blood is experiencing. For example, blood behaves as a Newtonian fluid in regions with a high shear rate when the blood cells undergo an intense mixing and average friction is not directly influenced by the corpuscular structure. Conversely, at a low shear rate, the collisions between individual cells give rise to higher friction and higher apparent viscosity. In the limit of very small shear, an effort is required to break the pre-existing pattern of red blood cells; a behavior that is sometimes modeled by a static yield stress. Viscosity is also a function of the local hematocrit because a higher percentage of red blood cells reflects into higher average friction.

Such variability of apparent viscosity is further influenced by several concurring factors and still lacks a comprehensive and satisfactory description that is valid in general. Therefore, several non-Newtonian models have been proposed for various applications. However, in large vessels, shear rates are normally high and these viscosity variations are small; moreover, the mathematics would present a significant increase in complexity when accounting for a variable viscosity. Therefore, at least for flow in large vessels, blood is normally treated as a Newtonian fluid with constant viscosity, which can vary from 3 to $4 \times 10^3 \, \mathrm{Kg/m \cdot s}$, which is about three to four times greater than the viscosity of water at $20°$.

Dimensionally, the dynamic viscosity is a proportionality coefficient expressed by (1.9) between a dynamic quantity, the shear stress that involves the three dimensional units (mass, length, time) and the shear rate that involves only kinematic units (length and time). When the left side of Eq. (1.9) is normalized with the density, the mass unit cancels and the same relationship can be rewritten in terms of kinematic quantities only as

$$\frac{\tau}{\rho} = \nu \frac{dv_x}{dy}, \; \nu = \frac{\mu}{\rho}, \tag{1.10}$$

which introduces the "kinematic viscosity" ν. The kinematic viscosity represents the viscosity coefficient directly involved in the description of fluid motion, whereas the dynamic viscosity enters when such motion must be translated into dynamic actions like forces and stresses. Kinematic viscosity takes value about $\nu = 10^{-6} \, \mathrm{m^2/s}$ for water (at $20°$) and $\nu = 1.5 \times 10^{-5} \, \mathrm{m^2/s}$ for air (at $20°$ and atmospheric pressure) showing that the motion of water is less viscous than air's although the involved shear stresses are larger because water density is larger. The value of kinematic viscosity for blood when assumed as a Newtonian fluid is $\nu \cong 3.5 \times 10^{-6} \, \mathrm{m^2/s}$, that is, more commonly expressed as $\nu \cong 3.5 \times 10^{-2} \, \mathrm{cm^2/s}$.

It must be clear in mind that the classification of materials as fluids and solids, as discussed above with their related properties, does not correspond to the inner nature of reality itself. Fluids and solids represent interpretative schemes, conceptual models to describe the behavior of reality under specific conditions. In a more profound perspective, the distinction between fluids and solids is not always immediate. Most materials present a simultaneous presence of both elastic and viscous behaviors. Some materials can be intrinsically viscoelastic (for example, gels). Some materials should be even be described as fluids in some conditions and as solids in others. A glacier is a solid if one can walk on it, yet it flows like a fluid during its slow motion detectable over the years. It is thus important to remind that solids and fluids, elasticity and viscosity, are conceptual models used to describe the behavior of specific materials within the limit of the specific situation of interest where these models may be appropriate.

That said, the fluid model is extremely accurate to describe the behavior of water, air, and similar elements in their liquid or gaseous phase; indeed, most related theoretical and applied advancements in fundamental physics, technology, and environmental studies are based on this scheme and its mathematical foundations.

1.3   Overview of Bio-Flow Domains

The ultimate goal of this book is a rigorous application of the principles of fluid dynamics to blood flow in the cardiovascular system. For completeness, we include here a quick overview of the circulatory system for the inexperienced reader. This is intentionally provided at an extremely superficial level; thus, the reader is redirected to the numerous other texts on the subject for more comprehensive descriptions.

Circulation is a system aimed to distribute nutrients (mainly oxygen) transported by the blood to every single cell. Reaching all regions in the body is a difficult task that requires efficiency both in the large vessels and in the microscopic intercellular space; to this aim, circulation uses two main mechanisms: transport and diffusion. Transport allows covering relatively large distances, from the heart to other body regions up to limbs. Blood is transported with the local velocity, say U, along the cardiovascular network and allow traveling a distance $\ell_{transp} \sim Ut$ in a time interval t. This mechanism is efficient until the velocity is high enough and becomes progressively less efficient in small vessels where velocity is much smaller. Indeed, velocity necessarily decreases at smaller scales in order to avoid the development of excessive shear stresses that are proportional to U/d, with d the vessel diameter, as shown by relationship (1.9). On the opposite end, when velocity is close to zero, diffusion becomes more efficient to cover small distances and permits the local distribution from capillary to interstitial space up to individual cells through a diffusive behavior that rapidly covers small distances. The length covered by diffusion in a time t can be estimated as $\ell_{diff} \sim \sqrt{2\nu t}$. Comparative results, reported in the table below, show how transport is well suited to cover large distances traveling along large vessels where velocity can be of the order of cm/s.

t	ℓ_{transp} ($U = 10$ cm/s)	ℓ_{transp} ($U = 1$ mm/s)	ℓ_{diff} ($\nu = 0.04$ cm^2/s)
10^{-3} s	0.1 mm	1 μm	90 μm
10^{-2} s	1 mm	10 μm	0.28 mm

(continued)

(continued)

t	ℓ_{transp} ($U = 10$ cm/s)	ℓ_{transp} ($U = 1$ mm/s)	ℓ_{diff} ($\nu = 0.04$ cm^2/s)
10^{-1} s	1 cm	0.1 mm	0.9 mm
1 s	10 cm	1 mm	2.8 mm
1 min	6 m	6 cm	2.2 cm
1 h	360 m	3.6 m	17 cm

When the vessels become small and velocities are reduced to values of few mm/s or smaller, diffusion becomes progressively more efficient to cover small distances.

The entire circulatory system is composed of the systemic and pulmonary circulation systems that are arranged in series. Figure 1.6 shows a sketch of the main vessels. Systemic circulation starts from the left heart, which receives low-pressure oxygenated blood from the pulmonary veins and pushes it at a higher pressure in the aorta, the first systemic artery. Aorta branches into smaller arteries that transport blood into different regions of the body, these in turn branch into smaller arteries then to arterioles and into capillaries that are close enough to any cell of the body to which oxygen is delivered and cells' refuses collected. Capillaries then merge into venules that merge further into progressively larger veins up to inferior vena cava and superior vena cava (from the lower and upper parts of the body, respectively) that end the systemic circulation and enter into the right heart. From the right heart, blood is pushed into the pulmonary arteries and branches up to the capillaries that go across the lungs, where red blood cells leave the refuses and collect oxygen. The pulmonary venous system downstream the lungs transport oxygenated blood and reaches the left heart to restart the cycle. The two circulations also operate in parallel because the left and right sides of the heart are part of the same organ and work in synergy with a common cardiac pace.

Mechanical analysis is principally dedicated to the transport mechanism inside large blood vessels, across the heart and along main arteries and veins, which also represent the sites of major clinical interest. There are important differences between arterial and venous networks. Blood flows in arteries through an unsteady, pulsatile motion forced by the heartbeat rhythm and fills arteries at high pressure (75 to 120 mmHg, that can be expressed as 1.0 to 1.6 $\times$ 10^{-5} Pa or as 1 to 1.6 mH$_2$O, thus blood may jump this high when an artery is punched). Differently, blood reaches the venous system after having passed through the capillary bed; there blood experienced large frictional resistances, it loses its unsteadiness and loses pressure. Thus, the venous flow is essentially a steady one and pressure is low (as immediately verifiable by pushing the superficial veins). This is also a reason why arteries have thicker walls and are protected deeper in the body, while veins are safely closer to the surface.

The diameter of arteries of higher pathophysiological interest range from few centimeters (aorta) to one centimeter or several millimeters (carotid bifurcation, iliac arteries), where unsteady velocities reach peaks of about 1 m/s or more. Fluid dynamics phenomena that are relevant to blood flow in the heart chambers and in the main arteries represent the main topics covered in later chapters.

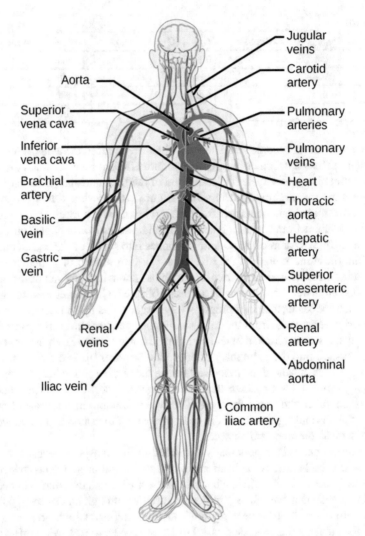

Fig. 1.6 Overview of the circulatory anatomy. The arterial system of the primary circulation, indicated in red, originates from the left side of the heart with the aortic arch and progressively branches to supply oxygenated blood to all the organs of the body. The corresponding venous system, in blue, collects de-oxygenated blood and returns to the right heart. In the pulmonary circulation, the arteries, in blue, start poor in oxygen from the right heart and the pulmonary veins with oxygenated blood end at the left side (credit: modification of work by Mariana Ruiz Villareal; CC BY)

This book will focus on blood flow in the large vessels of the cardiovascular system due to its paramount relevance with respect to other potential clinical application of fluid dynamics. Nevertheless, microcirculation as well as other aspects of biological fluid dynamics are gaining increasing attention for their potential significance to clinical applications.

Pulmonary ventilation system deals with the forced oscillatory motion of air across the pulmonary airways to the pulmonary alveoli. The main issues are the presence of dysfunctional/insufficient alveoli or the collapse of air vessels under extreme thrusts. In the same field, some attention is devoted to the fluid mechanics of main external airways (nasal sinuses, turbinate) for the numerous and common pathologies that affect these areas.

Biomechanics of the eye received particular attention during the past years because the eye is an organ composed of a series of chambers filled with biological fluids. The aqueous humor is a water-like fluid contained in the anterior and posterior chambers of the eye. It provides nutrients to the cornea and lens and regulates the intraocular pressure. The balance between aqueous production and resistance during drainage is an element participating in healthy eye function or to the development of diseases like glaucoma. The vitreous humor is a clear gel-like fluid that fills the vitreous chamber, between the lens and the retina. Its slow dynamics during eye movements is part of the normal ocular function, while an alteration of its composition and dynamics directly affects the retina. Numerous studies on theoretical aspects have created a firm ground for supporting actual clinical applications in the coming years.

Blood perfusion represents the small-scale dynamics of blood across the capillary bed inside the organs. There range from the liver to the kidney, to the muscles, up to the myocardial muscle. Perfusion analysis can be of interest to recognize and assess regions with insufficient blood supply (ischemic areas) following injury or a disease, or to assess potential absorption of drugs. Models of such systems are still at early stages for clinical application. More importantly, dedicated clinical imaging modalities allow direct evaluation of perfusion levels in numerous organs.

Industrial fluid dynamics is an essential part of clinical or biotechnological environments, in laboratories or plants, in either large or microscopic domains. It is, therefore, important to know the fundamental aspects of fluid dynamics to understand the main phenomena that allow the function of such systems.

These topics are not covered here, and the reader is redirected to other books for a wider spectrum of fluid mechanics phenomena that are present in biological environments (Fung, 1997; Rubenstein et al., 2015).

1.4 Dimensional Analysis

Before starting to talk about physical laws, it is important to dedicate some space to outline the fundamental topic of dimensional analysis. Dimensional analysis explores the implication of dimensional congruence for physical laws, and it is interesting to notice how this apparently trivial consideration can sometimes allow simplifying or even uncovering relationships between physical properties. However, this subject is more powerful than what is briefly described here and the reader is redirected to specific books for additional insights (Barenblatt, 2003). The topic of dimensional analysis is briefly introduced here with the aim of placing dimensional congruence at the basis of the physical descriptions presented later in this book and to provide an effective tool that will be employed in a few situations.

To this aim, let us start by the simple consideration that any physical property can be expressed in general as the product between a pure number and a dimensional unit. To be explicit, a property X can be expressed as $X = A \times$ UNIT$_1$ or $X = B \times$ UNIT$_2$. The usage of different units brings to a different numerical coefficient, which can be A times the UNIT$_1$ or B times the UNIT$_2$, however, the physical property itself is evidently not affected by a change in the unit used for its description. For example, a person height $X = 1.80$ m can be expressed as $X = 180$ cm or as $X = 70.87$ inch but the physical property itself, the height of that person, is evidently independent of the unit chosen to describe it.

Similarly, a "physical law" reflects a physical phenomenon that is independent of the units used to describe it. As before, this is a trivial affirmation; however, this simple concept is a constraint that allows simplification of the expression of the physical laws itself.

Let us use one example to show the power of dimensional analysis. Consider a fluid that flows inside a cylindrical vessel, the fluid moves because it is pushed upstream by a pressure (potential energy) that is larger than the value downstream thus giving a net propulsive force that overcomes the viscous friction experienced by fluid along its motion. Suppose that we are willing to express how the reduction of pressure per unit length depends on the parameters that are involved in the physical systems. Without any knowledge of the physical laws governing fluid dynamics, we can state that this phenomenon must be expressed by a physical law of the type

$$\frac{dp}{dx} = f(D, U, \rho, \mu), \tag{1.11}$$

which states that the pressure gradient (pressure loss per unit length) is a function of all the properties that may influence it: the vessel diameter D, the flow velocity U, and the fluid characteristics, density ρ and viscosity μ. Assuming a simple configuration (cylindrical vessel with no bend, obstacles, etc.), there are no other quantities coming into play. Thus, a physical law (1.11) must exist although its specific form may be unknown. This law (1.11) depends on 4 parameters, if you had to find its expression by performing a series of experiments, considering to test (as a minimum) 10 values for each parameter, you had to make 10^4 experiments to fill this 4-dimensional parameters' space. Requiring multiple experimental apparatuses with different diameters D and different fluids to vary ρ and μ.

Here is where dimensional analysis can help. Equation (1.11) is a physical law, thus it does not depend on the specific unit that we decide to use for the length, say L, time, T, and mass, M, chosen to express it. You can choose standard units

(L = m, T = sec, M = Kg) or Anglo-Saxon units (L = ft, T = sec, M = lb) or any other one; the resulting law would be unaffected by this choice. Once the units are decided, the physical law will express a relationship between the numerical coefficients expressing the quantities in those units and the law is automatically consistent because a physical law is independent of the choice of units. This concept can be stated more simply by requiring that units on the left and the right side of (1.11) must be the same.

However, it is not necessary to use units previously defined by some international standard in a separate context. It is actually smarter to use units that are natural to the specific application. In this case, one could use the diameter D as a unit of length, the ratio D/U as the unit of time and ρD^3 as the unit of mass

$$L = D, \ T = \frac{D}{U}, \ M = \rho D^3 . \tag{1.12}$$

Even with this special choice, the physical law will express a relationship between the numerical coefficients of every quantity given in those units. Thus, express each quantity in (1.11) as the product between the numerical coefficients and its units (1.12). The numerical coefficient is trivially obtained by dividing the dimensional quantity by its units. Thus, the gradient of pressure can be expressed as

$$\frac{dp}{dx} = \frac{D}{\rho U^2} \frac{dp}{dx} \cdot \left[\frac{M}{L^2 T^2} \right],$$

where $\frac{D}{\rho U^2} \frac{dp}{dx}$ is the numerical coefficient and $\left[\frac{M}{L^2 T^2} \right]$ is the unit, that is equal to $\frac{\rho U^2}{D}$. Application of the same concept to the other quantities gives

$$D = 1 \cdot [L], \quad U = 1 \cdot \left[\frac{L}{T} \right], \quad \rho = 1 \cdot \left[\frac{M}{L^3} \right], \quad \mu = \frac{\mu}{\rho U D} \cdot \left[\frac{M}{LT} \right]$$

Then insert these into (1.11) to obtain the relationship between the numerical coefficients because the units are automatically satisfied being it a physical law

$$\frac{D}{\rho U^2} \frac{dp}{dx} = f\left(1, 1, 1, \frac{\mu}{\rho U D}\right) = f\left(\frac{\mu}{\rho U D}\right). \tag{1.13}$$

Equation (1.13) represents the same physical law (1.11), but it is now stated as a relationship between dimensionless quantities. Expressed this way, the number of independent variables is reduced from 4 to a single one. Thus, you can establish the physical law by making just N experiments instead of N^4 that could even be performed, for example, just with one fluid in one vessel and varying the fluid velocity only.

This simplification allowed by dimensional congruence is a general rule: expressing a physical law in dimensionless terms allows reducing the number of variables by a number equal to the number of independent dimensional units involved in the law. In the previous case, a relationship between 5 variables involving 3 units has been simplified in a relationship between 2 dimensionless variables.

It is easy to demonstrate that the resulting law is independent of the specific units selected. In the previous example, we could use, for example, a different unit time unit

$$L = D \quad T = \frac{\rho D^2}{\mu} \quad M = \rho D^3.$$

These correspond to a different set of numerical coefficients

$$\frac{dp}{dx} = \frac{\rho D^3}{\mu^2} \frac{dp}{dx} \cdot \left[\frac{M}{L^2 T^2} \right], \quad D = 1 \cdot [L], \quad U = \frac{\rho D U}{\mu} \cdot \left[\frac{L}{T} \right],$$

$$\rho = 1 \cdot \left[\frac{M}{L^3} \right], \quad \mu = 1 \cdot \left[\frac{M}{LT} \right];$$

that inserted into (1.11) give

$$\frac{\rho D^3}{\mu^2} \frac{dp}{dx} = f \left(1, \frac{\rho D U}{\mu}, 1, 1 \right).$$

This relationship is equivalent to (1.13), because it can be recast as

$$\frac{D}{\rho U^2} \frac{dp}{dx} = \left(\frac{\rho U D}{\mu} \right)^{-2} f \left(\frac{\rho D U}{\mu} \right) = f \left(\frac{\mu}{\rho D U} \right),$$

to give a result functionally identical to (1.13).

It is actually a general result for fluid flowing in smooth cylindrical vessels that the pressure loss per unit of length is expressed in general as

$$\frac{dp}{dx} = \frac{\rho U^2}{2D} f(Re), \tag{1.14}$$

where f is known as the (Darcy) friction factor that depends on the ratio $Re = \frac{UD}{\nu}$, an important dimensionless number that is known as the Reynolds number and that we will encounter several times along this book. The Reynolds number represents the relative importance of kinetic energy with respect to viscous frictions; when Re is high the flow is vigorous and the relative entity of viscous friction is low, vice versa, a flow at low Re is slow and dominated by friction.

It is important to remark that the general resistance law (1.13) or (1.14) has been obtained based on dimensional consideration only, without using any knowledge of fluid dynamics. This example demonstrates the power of the simple concept of dimensional analysis. Fluid dynamics theory may be then advocated to better specify the function $f(Re)$; however, we will see that this is not immediate in most of the cases and (1.13) may become the only theoretical result to be integrated by physical experiments.

It is, therefore, of fundamental importance to formulate any physical law in dimensionless terms; this permits to have it expressed in the simplest way as possible and to identify the dimensionless parameters that describe the underlying physics.

If we extend the example (1.11) to consider a pulsatile flow with period T,

$$\frac{dp}{dx} = f(D, U, \rho, \mu, T). \tag{1.15}$$

Selection of the same units (1.12) gives the dimensionless relationship

$$\frac{D}{\rho U^2} \frac{dp}{dx} = f(Re, St) \tag{1.16}$$

showing that pressure changes as before due to friction (dependence on the Reynolds number, Re) and it also depends on the frequency of oscillation that is expressed by the Strouhal number $St = \frac{D}{UT}$.

Dimensional analysis permits to reduce the number of independent variables to their minimum and to recognize the dimensionless number that characterizes the phenomenon under analysis. It is a powerful tool when facing complex conditions, for example, when mathematical equations do not lead to a closed solution. It will be used in some occasions to progress across critical passages that cannot be solved otherwise.

Chapter 2
Fluid Statics

Abstract This chapter introduces the laws governing the transmission of forces across a fluid at rest. When the velocity is zero, the fluid exerts forces on the surrounding boundaries through the value of pressure in the contact regions. The laws of statics make it possible to derive the spatial distribution of the pressure field. The integration of the pressure field over a plane surface allows writing a simple rule to evaluate the force on such surface. Then, the concept is extended to curved surfaces to devise a general approach to compute the force made by a fluid onto the surfaces of arbitrary shape. A series of examples are provided to illustrate typical situations that can occur.

2.1 Pressure Distribution

Fluid statics deals with the forces transmitted by fluids in the absence of motion. These are of enormous importance in numerous applications, from industrial to biological, as they represent the basic stress state in every fluid domain. Motion, when it occurs, may induce modifications on top of this background static state.

Statics means that the velocity vector field is identically zero. As we have seen in the definition of fluids and of viscosity, shear stresses develop in consequence of differential velocities (shear rate, or rate of deformation). Therefore, in static conditions, shear stresses are also absent and the stress made by still fluid over any surface has only a normal component

$$\tau = p n, \tag{2.1}$$

where n is the normal to the surface (a vector perpendicular to the surface, directed toward the surface and of unit modulus) and $p(x)$ is the pressure field that can vary at different spatial positions indicated by the position vector x. It is important to remark from its first appearance that pressure is a scalar quantity. As such it has no direction; pressure gives rise to a stress vector (2.1) only after it acts on a surface in which case the direction is the normal n that is given by the orientation of the surface, directed toward it.

© The Author(s), under exclusive license to Springer Nature Switzerland AG 2022
G. Pedrizzetti, *Fluid Mechanics for Cardiovascular Engineering*,
https://doi.org/10.1007/978-3-030-85943-5_2

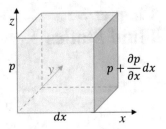

Fig. 2.1 Balance applied on infinitesimal cube

Statics obey the law of equilibrium, which states that the sum of all forces acting on a volume V of fluid must be zero

$$\int_V f dV + \int_S p \mathbf{n} dS = 0. \tag{2.2}$$

The first integral represents the forces acting on the volume, and f is the volumetric force density (e.g., specific gravity), the second integral represents the forces acting on the external surface S that surrounds the volume. Equation (2.2) is the integral balance equation of fluid statics.

The balance equation can also be expressed in differential form. To this aim, consider that the integral Eq. (2.2) is valid for an arbitrary volume, then take the special case of an infinitesimal cube of size $dx \times dy \times dz$, as shown in Fig. 2.1 the balance (2.2) along the x-direction is made by the x-component of the volume force, f_x, acting on the volume and the pressure that acts only on the two faces of area $dydz$ with normal $\mathbf{n} = [1, 0, 0]$ and $\mathbf{n} = [-1, 0, 0]$

$$f_x dx dy dz + p dy dz - \left(p + \frac{\partial p}{\partial x} dx \right) dy dz = 0.$$

Simplification gives

$$f_x = \frac{\partial p}{\partial x}.$$

That can be written for all coordinates and recast in vector form to provide the general differential equation of fluid statics

$$f = \nabla p. \tag{2.3}$$

Equation (2.3) states that every fluid particle is in balance between the volumetric forces and the pressure gradient. It is immediate to verify that the result (2.3) could also be obtained directly from Eq. (2.2) by transforming the second terms therein into a volumetric integral and then extending the equality to the terms inside the integral for the arbitrariness of the volume (or using an infinitesimal volume). This would have required the application of the Gauss theorem that will be recalled later in Sect. 3.2.

The volumetric force of greatest practical interest for applications is the gravitational force. It can be expressed as

$$f = -\gamma k = -\nabla \gamma z, \tag{2.4}$$

where k is the unit vector directed upwards against gravity, and z is the corresponding direction. The specific gravity is $\gamma = \rho g$, where g is the module of gravity acceleration. In case of gravitational forces, Eq. (2.3) is made by the sum of two gradient terms and takes the special expression

$$\nabla(p + \gamma z) = 0. \tag{2.5}$$

Equation (2.5) states that in a fluid subjected to gravitational field only, pressure can only vary with the quote z, $p(z)$; it is constant on xy-planes at constant z and it increases linearly as the quote z decreases. The interpretation of Eq. (2.5) becomes more immediate with the introduction of a new quantity called the *static head* defined as

$$h = z + \frac{p}{\gamma}. \tag{2.6}$$

Using this definition, Eq. (2.5) expresses the first fundamental concept of fluid statics: *the static head remains constant inside a same fluid*; in other words, the value of the static head is a property of a volume of fluid and characterizes its potential energy (per unit of weight). The constancy of the static head allows evaluating the pressure difference between two points at different quote z inside the same fluid. Consider two arbitrary points "0" and "1", the constancy of the static head states that $h_0 = h_1$, which tells

$$p_0 + \gamma z_0 = p_1 + \gamma z_1. \tag{2.7}$$

Thus

$$p_1 - p_0 = \gamma(z_0 - z_1), \tag{2.8}$$

the pressure difference between two points is equal to the difference of quote multiplied by the specific gravity. Intuitively, one can consider that the pressure at the lower point is increased by the weight of the column of fluid above it.

It is common to rewrite (2.8) taking point "0" at a reference position, which is often at the free surface of a reservoir subjected to atmospheric pressure or it can be the level of the heart in the human body, and point "1" at a generic level z

$$p(z) = p_0 + \gamma(z_0 - z). \tag{2.9}$$

It is useful to define the depth $\zeta = z_0 - z$, such that pressure grows linearly with the depth ζ from the reference quote and rewrite (2.9) in terms of the ζ coordinate

$$p(\zeta) = p_0 + \gamma\zeta \tag{2.10}$$

showing explicitly the growth of pressure with depth.

It can be noticed that pressure entered in all previous equations in terms of pressure difference only, thus adding a constant value to all pressure values does not change the formulas. Indeed, the absolute value of pressure does not enter explicitly in the balances as a force is always created by the difference of pressure between two regions. In particular, the atmospheric pressure surrounds most systems of interest and represents the reference value in most situations; for example, it adds to blood pressure everywhere in the circulatory system. Therefore, it is common to use atmospheric pressure as the reference zero value, and express pressure as the difference relative to atmospheric pressure. In the frequent case of a reservoir with a free surface, one can state that the (relative) pressure is zero at the free surface and increases with depth

$$p(\zeta) = \gamma\zeta. \tag{2.11}$$

The pressure relative to the atmospheric values is sometimes called relative pressure, or gauge pressure, or simply pressure. With this representation, it is also immediately evident from (2.6) that the level of the free surface where pressure is zero represents the value of the static head for that fluid.

In the circulatory system, Eq. (2.10) tells that a person in the orthostatic position is subjected to a static pressure that at the lower limbs is about 100 mmHg higher than the value at the level of the heart and at the head is about 30 mmHg below. Whereas static pressure is balanced in the supine position. The value of static pressure is important for the phenomena like tissue perfusion as well as for the adaptation of tissues to this level of stress. The static pressure also represents the underlying distribution on top of which pressure values change due to phenomena associated with blood flowing, as well as to the behavior of the surrounding tissues.

Although pressure will be commonly considered relative to the atmospheric pressure, a quick remark is due about the situations when the absolute value of pressure becomes important. This is related to the fact that pressure is defined as a positive physical quantity, absolute pressure cannot be negative or relative pressure cannot be lower than $-p_{Atm}$. The calculations above treat pressure as a real number and do not include this constraint; therefore, when, for some reason, these give rive rise to values lower than this limit the calculations are not valid. When pressure approaches the zero value, some different phenomena develop, and they require a special treatment. For example, there is a limit to the height that a fluid can be lifted because pressure decreases with height but cannot decrease below a relative pressure value

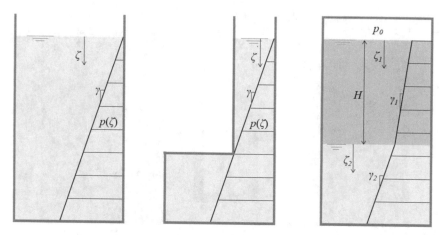

Fig. 2.2 Pressure distribution along the depth in a static fluid system

$p = -p_{Atm}$ (atmospheric pressure is about 10^5 Pa, which corresponds to a fluid column of about 10 mH₂O or 750 mmHg). Other dynamic phenomena associated with fluid motion can bring pressure close to this limit, a common one is cavitation that can occur sporadically in circulation and will be mentioned in Chap. 13.

The second fundamental concept of fluid statics is derived from the equilibrium of the interface between two fluids. Assuming that surface tension is negligible (in absence of capillary phenomena that should be treated separately) an interface between two fluids is subjected only to pressure on the two faces and its equilibrium gives the result that *pressure is the same on the two sides at the interface between two fluids*. In other words, pressure is continuous at the interface between two fluids.

These two principles make possible the evaluation of pressure difference between all places in fluid-filled containers. Figure 2.2 shows on the left a simple case where pressure increases linearly with depth from the zero value at the surface as dictated by Eq. (2.11). Equation (2.11) states explicitly that the pressure value at the bottom of the container depends on its depth relative to the surface only; therefore, an identical distribution of pressure is found on the container shown in the central picture of Fig. 2.2. This picture is included to stress the point that pressure is caused by depth only and not by the weight of the fluid above; the difference of pressure between two points at different depths can still be interpreted as due to the weight of the column of fluid between them; however, this applies assuming that fluid is continuous between the two points and, if an obstacle would be present, the pressure difference is still given by the difference in depth only. Differently, the right side of Fig. 2.2 shows a pressure profile with two immiscible fluids with different specific gravity, γ_1 and $\gamma_2 > \gamma_1$, and a non-zero pressure value at the free surface due the presence of a pressurized gas. Notice that a gas can be often considered having a uniform pressure because the specific gravity is very small and the difference with the quote is negligible unless considering a very large difference in quote (like between atmospheric layers). It is easy to show that, with reference to symbols in the figure, pressure takes the value

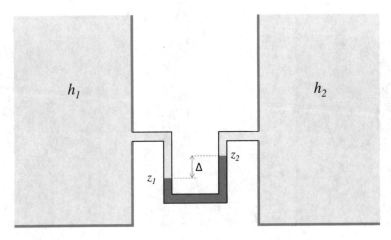

Fig. 2.3 Differential manometer

$p(\zeta_1) = p_0 + \gamma_1 \zeta_1$ in the upper, lighter fluid and $p(\zeta_2) = p_0 + \gamma_1 H + \gamma_2 \zeta_2$ in the lower one.

An interesting and technologically important case is the differential manometer schematically drawn in Fig. 2.3. It is a tool that allows measuring the difference in static head between two chambers filled with the same fluid of specific gravity γ connected by a small duct partially filled with a heavier fluid (typically mercury) of specific gravity γ_m. The static head is constant inside each reservoir; thus, we can use the points at the edge with the heavier fluid and, with reference to Fig. 2.3, using the definition of static head, write

$$h_1 - h_2 = \left(z_1 + \frac{p_1}{\gamma}\right) - \left(z_2 + \frac{p_2}{\gamma}\right) = \frac{p_1 - p_2}{\gamma} - \Delta.$$

Now apply the conservation of h inside the heavier fluid

$$z_1 + \tfrac{p_1}{\gamma_m} = z_2 + \tfrac{p_2}{\gamma_m}, \Rightarrow p_1 - p_2 = \gamma_m \Delta.$$

Substitution of $p_1 - p_2$ back into the previous formula gives the sought result

$$h_1 - h_2 = \left(\frac{\gamma_m - \gamma}{\gamma}\right) \Delta. \tag{2.12}$$

Equation (2.12) permits to compute the difference of static head between the two chambers from the reading of the difference of height Δ in the differential manometer. Often one chamber has a known head (for example, it is open to the atmosphere, or it is a pressurized gas) and it is used as a reference to measure directly the head in the other chamber. Once the head is known, pressure can be obtained at every point at quote z by the definition of static head (2.6).

The ability to know the pressure field inside a fluid takes on particular importance when evaluating the forces exerted by this on the surrounding boundaries. The pressure distribution will be used in the following sections to calculate the forces on surfaces in contact with fluids.

2.2 Forces on Plane Surfaces

Consider a planar surface with area A and normal $\boldsymbol{n}$ directed towards the surface. The force vector acting on the surface is by definition the integral of stresses, $\boldsymbol{\tau} = p\boldsymbol{n}$, on it

$$F = \int_A pn\,dA = Fn. \tag{2.13}$$

The force vector can be expressed by the product of its modulus F and direction vector $\boldsymbol{n}$; which is constant because the surface is plane. Consider the surface wet by a single fluid whose pressure can be expressed in general by expression (2.9)

$$F = \int_A p(z)dA = p_0 A + \gamma z_0 A - \gamma \int_A z\,dA = p_0 A + \gamma(z_0 - z_G)A = p_G A,$$

where we have used the general definition for the geometric center G of a surface

$$x_G = \frac{1}{A}\int_A x\,dA \tag{2.14}$$

applied here to the z coordinate. Thus, the force made by a fluid on a plane surface has always modulus equal to the pressure on the center of the surface multiplied by the area of the surface, and it is directed toward the surface

$$F = p_G A\boldsymbol{n}. \tag{2.15}$$

The force (2.15) is the integral result of the distribution of pressure over the surface. Although the force is computed by the value of pressure in G, let's remark with emphasis that the force is applied in a point P that is not the center G of the surface. The point of application is the center of the pressure distribution that is usually below the center of the surface because pressure is higher at a higher depth.

The knowledge of the point of application P is needed to compute the torque T caused by the force for the rotational equilibrium about an axis. To this aim, consider the simple case of a vertical surface allowed to rotate about a horizontal axis lying on the surface at depth ζ_1. By definition, the torque is given by the integral of the infinitesimal torques made at every depth ζ by the infinitesimal force $p(\zeta)dA$ with arm $(\zeta - \zeta_1)$

$$T = \int_A (\zeta - \zeta_1)p(\zeta)dA = (\zeta_P - \zeta_1)F, \tag{2.16}$$

which must be equivalent to the torque given by the force F applied at the depth ζ_P of the point of application. It is convenient to measure the depth with reference to the level where $p = 0$; in that case, it is immediate to show that the depth of the point of application is

$$\zeta_P = \frac{\int_A \zeta^2 dA}{\int_A \zeta dA}. \tag{2.17}$$

It is also immediate to verify that the result (2.17) remains valid when the surface is not vertical. The integrals in Eq. (2.17) are easy to evaluate when the surface is a rectangle contained between two depths ζ_A and $\zeta_B > \zeta_A$

$$\zeta_P = \frac{2}{3} \frac{\zeta_B^3 - \zeta_A^3}{\zeta_B^2 - \zeta_A^2}. \tag{2.18}$$

In the limiting case when the surface's upper edge is on the free surface, $\zeta_A = 0$, the distribution of pressure is triangular and the center of pressure P is at a depth 2/3 the surface height. In the other limit, in general, when the surface is horizontal, pressure is constant and the center of the pressure distribution coincides with the geometric center of the surface $P = G$.

Rectangular surfaces are the most common; in this case, it is sometimes convenient to divide the pressure distribution as the sum of a rectangular profile, applied in the surface center, and a triangular profile applied at two-thirds the depth; then compute the torque as the sum of the two individual ones. In more complex surfaces, the torque can be computed by dividing it into a composition of simpler surfaces.

As an instructive example, with reference to Fig. 2.4, consider the plane surface hinged at a generic point C, that separates two chambers containing the same fluid, one with level HA and the other with a higher level HB.

The torque made by the fluid can be computed separately on the left and right sides of the surface by evaluating the forces and the corresponding arms. On the left side, the distribution of pressure is triangular and the force is applied at two-thirds of the depth. On the right side, the pressure distribution is trapezoidal and it is useful to separate it as the sum of a rectangular and a triangular part. Reminding that each

Fig. 2.4 Calculation of the torque on an inclined surface separating two chambers with different static height

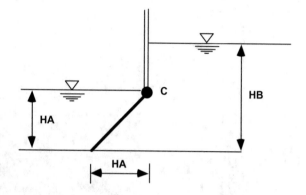

force acts perpendicular to the inclined surface, and assuming positive the actions on the right side (clockwise rotation), we get (indicating with subscripts L and R the forces on the left and right sides, respectively)

$$F_L = -\gamma \mathrm{HA}^2 \tfrac{\sqrt{2}}{2}, \; r = \tfrac{2\sqrt{2}}{3}\mathrm{HA}, \; T_L = -\gamma \tfrac{2}{3}\mathrm{HA}^2;$$
$$F_{R1} = \gamma(\mathrm{HB} - \mathrm{HA})\mathrm{HA}\sqrt{2}, \; r = \tfrac{\sqrt{2}}{2}\mathrm{HA}, \; T_{R1} = \gamma(\mathrm{HB} - \mathrm{HA})\mathrm{HA}^2;$$
$$F_{R2} = \gamma \mathrm{HA}^2 \tfrac{\sqrt{2}}{2}, \; r = \tfrac{2\sqrt{2}}{3}\mathrm{HA}, \; T_{R2} = \gamma \tfrac{2}{3}\mathrm{HA}^2.$$

The total torque is then given by the summations of the individual contributions. One can notice, however, that F_L and F_{R2} cancel each other exactly. Indeed, the triangular distribution of pressure on the left is equal and opposite to that on the right side; one could have noticed from the beginning that the net distribution of the pressure, canceling the opposite contributions coming from the two sides, is a rectangular distribution whose torque is T_{R1}.

When the distribution of pressure is rectangular, the fluid problem is reduced to that of finding pressure at a single point. In the previous example, that point was on the right side at a depth HB $-$ HA. This is particularly clear when a surface is placed horizontally because pressure takes the same value at all points. For example, consider the system in Fig. 2.5 where we want to compute the force and the torque acting on the square cover of side length d. Consider the specific weigh of fluid γ, and that of mercury γ_m in the differential manometer whose reading is Δ.

If the static height in the external chamber is h_1, which in the main chamber can be evaluated from the reading in the differential manometer

$$h = h_1 - \left(\tfrac{\gamma_m - \gamma}{\gamma}\right)\Delta.$$

Thus, the second chamber is like having a free surface placed at a height h and it is immediate to compute the pressure and the torque on the square surface as

$$F = \gamma h d^2, \; T = F\tfrac{d}{2}.$$

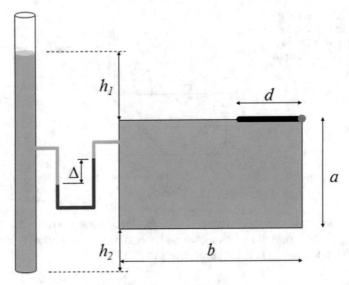

Fig. 2.5 Static fluid system with a flat square cover in a closed chamber connected with a differential manometer to an external reservoir

As a further example, let us try to evaluate the width X of the base of the two connected surfaces in.

Figure 2.6 such that they are in equilibrium to overturning.

Equilibrium to rotation tells that the torque on the inclined surface must be equal to that on the horizontal surface. This gives

$$\gamma \frac{hL}{2} \frac{L}{3} = \gamma h X \frac{X}{2}, \ X = \frac{L}{\sqrt{3}}.$$

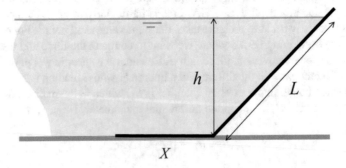

Fig. 2.6 Calculation of forces and torque acting on two connected flat surfaces

All these methods provide an immediate understanding of the expected results in those simple cases that are most frequent. They are also useful for drawing approximate results in complex conditions. In general geometries, the integrals in (2.13) and in (2.16) can be computed with ease by numerical integration.

2.3   Forces on Curved Surfaces

Consider now a generic surface S, with an arbitrary curved shape. An infinitesimal force $d\mathbf{F} = p\mathbf{n}dS$ acts at every individual infinitesimal element of surface dS directed with the local normal $\mathbf{n}$ that varies at the different points on the surface. The total force acting on S is given by the surface integral of all such infinitesimal forces

$$\mathbf{F} = \int_S d\mathbf{F} = \int_S p\mathbf{n}dS. \tag{2.19}$$

Differently from the case of plane surfaces, the normal $\mathbf{n}$ is not a constant and the integral cannot be simplified like it was done therein. A method to compute (2.19) can be obtained by advocating the global balance (2.2).

Consider first the case of a *closed surface*: a surface S surrounding a volume V that is in equilibrium immersed in a fluid. The force is given by the integral (2.19) evaluated on the external side of the surface S. It is important to remind that the value of the integral depends on pressure values in the fluid on the external edge of the volume V; therefore, it is independent on whether the inside volume V is occupied by a body (kept static by some mean) or it is a volume of fluid, because under static conditions the distribution of pressure depends on the depth of each point only. Consider first the case where V is of a volume of fluid and S is thus a mathematical surface with the same fluid on both sides. We are under statics condition and the volume of fluid V is in equilibrium; this means that the sum of all forces acting on the volume is zero. Following the integral balance (2.2), these forces are composed by the weight of the fluid volume V and the integral of pressure on the surface S

$$-\gamma V\mathbf{k} + \int_S p\mathbf{n}dS = 0 \tag{2.20}$$

with $\mathbf{k}$ the unit vector directed against gravity. The second term in (2.20) is the force (2.19) made by the fluid on the external surface of that volume, thus equilibrium tells that the horizontal components of the forces are zeros and vertical force is directed upward (buoyancy force). In summary, the force acting on an arbitrary closed surface surrounding a volume V is given by

$$\mathbf{F} = \gamma V\mathbf{k}. \tag{2.21}$$

This result does not depend on whether the surface surrounds a volume V of fluid or it is the external surface of a solid body of the same volume kept in equilibrium by some external mean. The force made by the fluid on the surface of the body is given by the same integral (2.21), because pressure distribution and the value of that integral are independent of the presence of the body. Equation (2.21) states Archimedes' principle (dated back to the third century BC), which tells that "a body immersed in a fluid is subjected to a force directed upward that is equal to the weight of the displaced fluid".

Consider the overall force acting on a solid body with its own weight $\gamma_s V$, being γ_s the solid specific gravity; when it is immerged in a fluid, the body is subjected to its weight and the buoyancy force (2.21). As a result, the apparent weight of a body immersed in a fluid is reduced by buoyancy and becomes $(\gamma_s - \gamma)V$. The value $(\gamma_s - \gamma)$ represents the apparent specific gravity of an immersed body and it is often useful for immediate evaluations.

Let us now move on and consider a generic surface S that can be open and have fluid on one or the other side. The procedure to compute the force acting on the surface S, Eq. (2.19), is that of selecting a volume of fluid partly surrounded by S and partly closed by planar surfaces. That volume is in equilibrium and obeys the law (2.2) that represents a balance of forces. These forces comprise volumetric forces that can be calculated, forces on plane surfaces that we have learned above how to calculate, and the force on the curved surface that remains the only unknown in the balance.

This apparently complex procedure is relatively straightforward in practice. Consider, for example, a surface made of a quarter of a circumference like the one shown in Fig. 2.7.

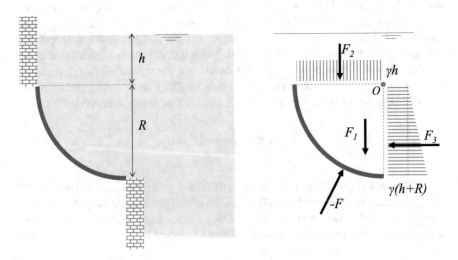

Fig. 2.7 Calculation of the force on a curved surface made of a quarter of a cylinder

Select an arbitrary control volume to perform the balance of forces, for example, the quarter of cylinder of radius R. The forces (per unit width) acting on that volume are.

- the weight of the fluid volume $F_1 = \gamma V$ directed downward;
- the force on the horizontal plane boundary F_2 where pressure is γh and the normal is directed downward;
- the force on the vertical boundary F_3 where pressure varies linearly from γh to $\gamma(h + R)$;
- the unknown force vector made by the curve surface to the fluid volume, that is equal in module and opposite in direction to the force $F = [F_x, F_y]$ made by the fluid on the surface.

Assuming the force directed as in in Fig. 2.7, the balance along the horizontal rightward direction, x, and vertical upward direction, z, gives

$$F_x = -F_3 = -\gamma\left(h + \tfrac{R}{2}\right)R$$
$$F_z = -F_1 - F_2 = -\gamma\tfrac{\pi}{4}R^2 - \gamma h R$$

showing that the force on the surface is directed leftward and downward.

The result is always independent of the chosen volume, although some choices permit an easier calculation. For example, in this case, it is immediate to see that the same result would have been achieved by selecting a volume extending up to the free surface. In that case, the force F_2 would be substituted by the weight of the added volume of fluid and the horizontal forces on the two sides of that adder volume would be identical and opposite. We could also use a smaller volume bounded by the chord connecting the two extremes of the surface; calculations are less immediate, but the result is identical.

It is useful to always keep in mind that the force is the integral of pressure on the surface as defined by (2.19), and that such calculation must not necessarily be performed on a fluid volume that is effectively present in the current configuration. It is, therefore, advisable to idealize the problem under investigation by extracting the mathematical surface and ideally immersing it in an unbounded fluid at the same depth as the original configuration. It is then easier to select a volume bounded by such an ideal surface reminding that the integral of pressure on that ideal surface is identical to that in the original configuration. For example, in Fig. 2.7, if the fluid was on the other side of the surface we would consider the volume on the wet side given by a square minus a circle. But we could also consider exactly the same volume (a quarter of a circle) noticing that the distribution of pressure on the external face is identical and opposite to that on the internal face, because pressure depends on depth only. Thus, the modulus is the same, and the direction opposite; the calculation could be the same as before and simply changing the sign to the resulting force.

A balance like (2.2) can be extended to the moment of forces, simply multiplying each force with the corresponding arm, to evaluate the torque on a surface. The explicit formula is not reported for brevity as the concept is exactly analogous to what was previously described for the forces. Once the control volume is selected, the balance of moment of forces is made with torques acting on planar surfaces bounding the volume, which can be computed by Eq. (2.16) and the following concepts therein, the moment of each weight force is immediate to evaluate as it is applied at the center of mass of the volume, eventually the torque on the curved surface remains as the only unknown in the balance.

Let us apply this concept to the example in Fig. 2.7 to evaluate the moment of the force on the surface S relative to a center of rotation in O. The torque balance, assumed positive counterclockwise, can be written as

$$T_O = F_1 r_1 + F_2 r_2 - F_3 r_3 = \gamma \frac{\pi}{4} R^2 \cdot \frac{4R}{3\pi} + \gamma h R \cdot \frac{R}{2}$$
$$- \gamma \left(h + \frac{R}{2} \right) R \cdot \left(\frac{2}{3} \frac{(h+R)^3 - h^3}{(h+R)^2 - h^2} - h \right).$$

The arm r_1 of the weight is obtained knowing that the center of mass of half a circle is displaced by a length $r_1 = \frac{4}{3\pi} R$ with respect to the center, this can be evaluated by the definition of center of mass (2.14). The force F_2 is made by a uniform distribution of pressure and it is applied at its center, thus $r_2 = R/2$. The torque associated to the force F_3 is evaluated by knowing its center of application using the result (2.18). For the calculation of this last torque, we could use an alternative approach that is sometimes easier by dividing the force due to the trapezoidal distribution of pressure as the sum, say $F_3 = F_{3a} + F_{3b}$, of that of a rectangular distribution that gives $F_{3a} = \gamma h R$ and whose arm is in this case equal to $r_{3a} = R/2$, plus the triangular distribution $F_{3b} = \gamma R^2/2$ whose arm is equal to $r_{3b} = 2R/3$. In this case, the balance can be rewritten as

$$T_O = F_1 r_1 + F_2 r_2 - F_{3a} r_{3a} - F_{3b} r_{3b} = \gamma \frac{\pi}{4} R^2 \cdot \frac{4R}{3\pi} + \gamma h R \cdot \frac{R}{2}$$
$$- \gamma h R \cdot \frac{R}{2} - \gamma \frac{R^2}{2} \cdot \frac{2}{3} R.$$

It is immediate to verify that both formulas are equivalent, although the second was easier to formulate and compute. Both formulas give the same result that the torque in the surface S about the point O is exactly zero in this case. This result could be anticipated here because the curve is a portion of a cylinder and every individual (infinitesimal) force acting normally to its surface is directed along the radius and presents zero torque about the center O of the circumference.

Static forces represent the underlying interaction between fluid and surrounding boundaries in every system involving fluids. This fact makes them the background system of forces on top of which changes due to fluid motion develop. Therefore, it is of fundamental importance to have a clear feeling of the system of static fluid forces that establishes in various situations.

To this aim, we present below a few instructive examples where the static force can be readily computed by a simple application of the balance laws described above.

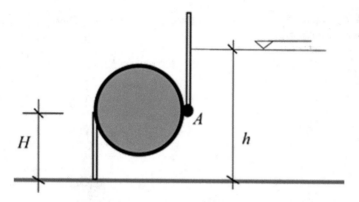

Fig. 2.8 Calculation of force and torque on a hinged cylinder

Consider the system sketched in Fig. 2.8 with a cylindrical body of radius R hinged in A. Here pressure acts from below due to the static height that is present on the right side. Therefore, we have only the force on a semicircle placed at a depth starting at $h - H$. The horizontal force is zero because, due to symmetry, this semicircle has no vertical surface closing it. The vertical force is given by the sum of the force of the flat horizontal surface plus the weight of the volume of fluid of the semicircle

$$F_V = \gamma\left(\tfrac{\pi}{2} R^2 + 2R(h - H)\right).$$

This force acts on the middle and gives a torque at the hinge

$$T_A = F_V \cdot R.$$

A similar approach can be used to compute the static force acting on the semi-spherical surface at the bottom of the chamber in Fig. 2.9. The force is identical to that of a half-sphere placed at the same depth in an unbounded fluid. The horizontal force is zero and the vertical force on the curved surface must balance the force on the flat circular surface at depth H, directed upward, and the weight of the half-sphere of fluid.

$$F = \gamma\left(\pi R^2 H - \tfrac{2}{3}\pi R^3\right).$$

Fig. 2.9 Calculation of the
static force on a
hemispherical surface at the
bottom of a chamber

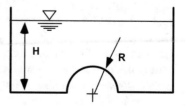

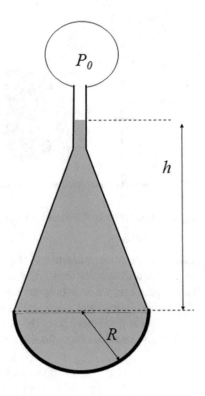

The force computed above can also be computed as the weight of the fluid above the surface. This interpretation is valid because the chamber extends up to the free surface. However, it must be kept in mind that forces are due to pressure, not to the weight of the fluid, and that pressure depends on depth only, independently from the geometry of the chamber. In order to clarify this point, let us consider the hemisphere at the base of the bowl shown in Fig. 2.10. The surface is placed at a depth h relative to the interface with the upper gas at a pressure P_0.

As before, the force is made by the force on the circular surface at depth h plus the weight of the half-sphere of fluid

$$F = (P_0 + \gamma h)\pi R^2 + \gamma \tfrac{2}{3}\pi R^3 .$$

This case underlines, besides the additional gas pressure, that the force is not related to the weight of the fluid above; it is rather the weight of the *hypothetical* fluid that would be above if the fluid extends straight above up to the free surface. The resulting force is in fact independent from the geometry of the bowl and from the inclination of its lateral surface. To understand this apparent contradiction, one must consider that there is a non-zero pressure on the inclined lateral surface that gives a downward-directed force that adds to the weight of the actual fluid; and

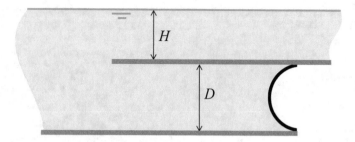

Fig. 2.11 Force on a hemispherical surface placed vertically

it is immediate to see that such additional force corresponds to the weight of the hypothetical fluid above such lateral surface.

As a final example, consider the force made by the fluid on the semispherical surface of diameter D placed vertically to close a chamber starting from a depth H. This is a set of general configurations of which one example is shown in Fig. 2.11. Identify a volume of fluid by closing the surface with a vertical circular flat surface. The horizontal force is that on the circular surface whose center is at depth $H + \frac{D}{2}$; the vertical force is simply the weight of the fluid volume

$$F_H = \gamma\left(H + \tfrac{D}{2}\right)\pi\tfrac{D^2}{4}, \; F_V = \gamma\tfrac{2}{3}\pi\tfrac{D^3}{8}.$$

A similar case can be taken when the sphere protrudes outward of the chamber instead of inward as in the previous example. In that case, the procedure for the calculation of the force would be quite same. The horizontal force is unchanged and the vertical force takes the same value but changes its sign.

Chapter 3
Fluid Kinematics

Abstract The description of complex fluid motion requires the definition of appropriate tools. This chapter recalls some basic notions of differential vector calculus that are extensively used afterward, paying special attention to the Gauss theorem that plays a central role in stating the equivalence between changes inside a volume and fluxes through its boundary. The act of fluid movement is divided into its elementary parts corresponding to different physical phenomena that will be later involved in conservation laws. Physical conservation laws express changes of properties associated with material objects; however, individual fluid elements cannot be followed during their mixing and swirling fluid motion beyond a very short time. This requires transferring all concepts in a different framework based on the properties evaluated on fixed points of space rather than on fixed material elements. All such content is presented with the aim to introduce a common mathematical notation and paying attention to accompany every mathematical concept with a physical, intuitive interpretation.

3.1 Recalls of Differential Vector Calculus

Let us first introduce the differential vector operator *nabla* that is useful to express derivatives in three-dimensional (3D) fields. In Cartesian coordinates, the operator nabla is defined as

$$\nabla = \begin{bmatrix} \frac{\partial}{\partial x} \\ \frac{\partial}{\partial y} \\ \frac{\partial}{\partial z} \end{bmatrix}.$$

The *gradient* of a scalar field $f(x)$ is a vector field, ∇f, obtained by applying the operator Nabla to it. In Cartesian coordinates, the gradient of a scalar field is a vector

© The Author(s), under exclusive license to Springer Nature Switzerland AG 2022
G. Pedrizzetti, *Fluid Mechanics for Cardiovascular Engineering*,
https://doi.org/10.1007/978-3-030-85943-5_3

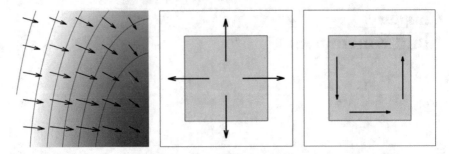

Fig. 3.1 Left: vector field given by the gradient of a scalar field (shown in level of brightness and level lines). Center: divergence about a point. Right: curl about a point

$$\nabla f = \begin{bmatrix} \frac{\partial f}{\partial x} \\ \frac{\partial f}{\partial y} \\ \frac{\partial f}{\partial z} \end{bmatrix} \tag{3.1}$$

that describes how the field f changes in space. For example, each component of ∇f tells how the field f changes along the corresponding direction. This concept can be restated that the gradient vector is always perpendicular to the lines or surfaces where f is constant, an example is shown in Fig. 3.1 (left). The gradient field is a vector and obeys the general rules of vector calculations; therefore, knowledge of the gradient vector permits the evaluation of partial derivatives along an arbitrary direction, say n, by standard vector projection

$$\frac{\partial f}{\partial n} = \nabla f \cdot n. \tag{3.2}$$

Gradient vector fields are important in physics. There are some situations when a physically relevant vector field, F, can be expressed as the gradient of a scalar field, $F = \nabla f$; in this case, the field F is a conservative field and the scalar field f is called the potential of F. It is immediate to verify that when $F = \nabla f$, its integral along a curve is trivially the difference of the potential f at the two ends and does not depend on the path itself

$$\int_a^b F \cdot ds = \int_a^b \nabla f \cdot ds = \int_a^b \frac{\partial f}{\partial s} ds = f_b - f_a.$$

This implies that the integral along any closed path is identically zero, which is the definition of a conservative field.

We have seen that the gradient of a scalar field is a vector field; indeed, the gradient operation increases the dimensionality. Similarly, the gradient of a vector field, say $v(x)$, is a tensor field, ∇v, whose component i, j in Cartesian coordinates is

$$(\nabla v)_{ij} = \frac{\partial v_i}{\partial x_j}. \tag{3.3}$$

The *divergence* of a vector field, $v(x)$, is a scalar field, $\nabla \cdot v$, obtained by performing formally the scalar product of v with the nabla operator. In Cartesian coordinates, the divergence is

$$\nabla \cdot v = \frac{\partial v_x}{\partial x} + \frac{\partial v_y}{\partial y} + \frac{\partial v_z}{\partial z} = \sum_{i=1}^{3} \frac{\partial v_i}{\partial x_i} = \frac{\partial v_i}{\partial x_i}, \tag{3.4}$$

where in the last equality the summation on repeated indices it is implicitly assumed (Einstein notation). The name divergence comes because a positive divergence at a point means that the vector (net of a constant value) is directed radially away from that point, thus it diverges. For example, consider the 2D case shown in Fig. 3.1 (center), $\frac{\partial v_x}{\partial x} + \frac{\partial v_y}{\partial y} > 0$ means that v_x is negative before and positive after a point and that v_y is negative below and positive above, thus the vector arrows depart away (diverge) from the point.

Vector fields with zero divergence are called *solenoidal* (name coming from electromagnetism) and take particular relevance in fluid dynamics as will be shown shortly.

The divergence reduces the dimensionality; thus, the divergence of a tensor field $\mathbb{T}$ is a vector field.

$$(\nabla \cdot \mathbb{T})_i = \frac{\partial \mathbb{T}_{ij}}{\partial x_j}. \tag{3.5}$$

Last vector operator is the *curl* that is applied to a vector field and produces another vector field; it does not change the dimensionality. The curl or a vector field is $\nabla \times v$, obtained by performing formally the internal product with nabla. In Cartesian coordinates, the curl is

$$\nabla \times v = \begin{bmatrix} \frac{\partial v_z}{\partial y} - \frac{\partial v_y}{\partial z} \\ \frac{\partial v_x}{\partial z} - \frac{\partial v_z}{\partial x} \\ \frac{\partial v_y}{\partial x} - \frac{\partial v_x}{\partial y} \end{bmatrix}. \tag{3.6}$$

When the vector field $v(x)$ has a positive curl, it means that it rotates counterclockwise about that point; indeed, when $v(x)$ is a velocity field, the value of $\nabla \times v$ represents twice the angular velocity. Figure 3.1 (right), shows the 2D case where only the third component $\frac{\partial v_y}{\partial x} - \frac{\partial v_x}{\partial y}$ is different from zero, the direction of the curl indicates the axis or rotation and its modulus the intensity of rotation.

The curl of velocity takes a special relevance in fluid dynamics and deserved its own name; the *vorticity vector field* $\omega(x)$

$$\boldsymbol{\omega}(\boldsymbol{x}) = \nabla \times \boldsymbol{v}, \tag{3.7}$$

which will be treated with attention in Chap. 10 to analyze the most advanced phenomena in cardiovascular flows. In particular, it will be shown therein that vector fields whose curl is zero are called *irrotational*; these are especially simple fields whose velocity can be expressed as a gradient of a potential.

As a final note, it is immediate to verify that the divergence of a curl field is identically zero (vorticity is a solenoidal field) and that the curl of a gradient is zero (conservative fields are irrotational fields)

$$\nabla \cdot (\nabla \times \boldsymbol{v}) = 0, \quad \nabla \times (\nabla f) = 0.$$

3.2   The Gauss Theorem in Integral Calculus

The Gauss theorem (or divergence theorem) states that the divergence of a vector field $\boldsymbol{v}(\boldsymbol{x})$ inside a volume V is equal to the flux of that vector across the boundary surface S of that volume

$$\int_V \nabla \cdot \boldsymbol{v} dV = \int_S \boldsymbol{v} \cdot \boldsymbol{n} dS, \tag{3.8}$$

where $\boldsymbol{n}$ is unit normal, directed outward. Despite the apparent mathematical complexity, the physical interpretation of the Gauss theorem is very intuitive. It can be seen immediately when the volume is a small cube or a square, in 2D, as depicted in Fig. 3.1 (middle picture): if the small region presents a divergence, this means that vector field points outward that cube. More in general, with reference to Fig. 3.2, when a vector field presents positive divergence inside a volume, it necessarily presents a net component directed outward of the bounding surface. On the contrary, when the total divergence is zero, the vector field can be directed outward in some regions and inward in some others but the integral over the bounding surface is zero because overall the vector field does not diverge.

The Gauss theorem permits to transform the calculation of a volume integral into a calculation on its boundary; in other words, it transforms a calculation based on volumetric, 3D information into another based on the information on the 2D bounding surface. Therefore, it represents a powerful mathematical tool in many contexts.

A simple application of the Gauss theorem permits to compute the volume of an arbitrary shape by the information on the geometry of its surface (Riley et al., 2006). For this, consider a field with unit divergence, $\nabla \cdot \boldsymbol{v} = 1$, for example, the field $\boldsymbol{v} = \boldsymbol{x}/3$ has unit divergence. Apply the Gauss theorem to this field to find that the volume can be computed by a surface integral

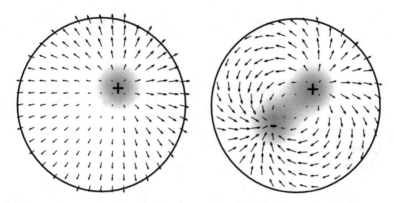

Fig. 3.2 Left: the presence of a positive divergence inside a volume corresponds to a vector field pointing outwards of the bounding surface. Right: when an opposite sign divergence is added to make the total divergence zero, some surface region presents outward pointing vectors some others inward vectors, with a net sum equal to zero

$$V = \int_V dV = \frac{1}{3} \int_S x \cdot n \, dS. \tag{3.9}$$

Similarly, the geometric center x_G of a volume can be computed as (Messner and Taylor, 1980)

$$x_G = \frac{1}{V} \int_V x \, dV = \frac{1}{4V} \int_S x(x \cdot n) \, dS. \tag{3.10}$$

The Gauss theorem can be rewritten for 2D space, transforming the volume in an area A and the surface in a closed curve C bounding the area.

$$\int_A \nabla \cdot v \, dA = \oint_C v \cdot n \, ds. \tag{3.11}$$

Equation (3.9) can be rewritten in 2D for computing the area of a generic figure that reads

$$A = \int_A dA = \frac{1}{2} \oint_C x \cdot n \, ds = \frac{1}{2} \oint_C (x \, dy - y \, dx), \tag{3.12}$$

where we used that $n \, ds = [dy, -dx]$.

The Gauss theorem can be rearranged to provide a number of different relationships between the volume and surface integral. A typical example is the integral of a curl

$$\int_V \nabla \times v dV = -\int_S v \times n dS, \tag{3.13}$$

which is obtained by applying (3.8) to a vector $u_j = \varepsilon_{ijk} v_k$, a, for an arbitrary value of $i = 1, 2, 3$ and repeating it for the three coordinates $i = 1, 2, 3$, (where ε_{ijk} is the fully antisymmetric tensor, or Levi–Civita tensor, equal to $+1$ when the three indices are a cyclic permutation of $1, 2, 3$, equal to -1 if an anticylcic permutation, and zero if two indices are equal.

Another example is the integral of a gradient field

$$\int_V \nabla f dV = \int_S f n dS \tag{3.14}$$

obtained as before by applying (3.8) to a vector $v_j = \delta_{ij} f$, for an arbitrary value of i, (where δ_{ij} is the identity tensor, equal to $+1$ when the two indices are equal and zero otherwise).

The Gauss theorem can also be used to compute the volume integral of a divergence-free vector field v that becomes

$$\int_V v dV = \int_S x(v \cdot n) dS. \tag{3.15}$$

This is obtained by applying (3.8) to the vector $u_j = v_j x_i$ as follows:

$$\int_V \nabla \cdot (v x_i) dV = \int_V (x_i \nabla \cdot v + v \cdot \nabla x_i) dV = \int_V v_i dV$$

$$= \int_S x_i v \cdot n dS$$

which gives (3.15) once repeated for the three coordinates $i = 1, 2, 3$ (Pedrizzetti, 2019).

Another fundamental theorem of integral calculus is the Stokes theorem (or circulation theorem) (Riley et al., 2006), which states that the circulation of a vector along a closed curve C is equal to the integral of its curl perpendicular any surface S bounded by that curve

$$\int_S (\nabla \times v) \cdot n dS = \oint_C v \cdot ds; \tag{3.16}$$

where the normal to the surface n and the direction of integration ds along the curve C are related by the right-hand rule. The Stokes theorem can also be derived from the Gauss theorem (and vice versa), in particular, it is a 2D version of the result (3.13), where the first term accounts for a single component of curl inside the surface S and the second term corresponds to integration along its boundary. The physical interpretation of the circulation theorem is similarly straightforward: the total rotation (circulation) around a closed curve is given by the summation of the individual rotations (curl of the vector) component contained inside that curve. Intuitively, one can refer to the simple case of a small square in 2D, as depicted in Fig. 3.1 (right picture), showing that a region with a non-zero curl necessarily corresponds to a circulating vector field. As a trivial example, take the rigid rotation of a circular plate with angular velocity Ω, the rotation velocity at a distance r from the center is $v_\theta = \Omega r$, the Stokes equality states that $2\pi v_\theta r = \pi r^2 \omega$, where ω is the component normal to the plane of the vorticity, previously introduced in (3.7), whose value is twice the angular velocity $\omega = 2\Omega$.

3.3 Breaking Down Elementary Motion

Consider the velocity $v(x)$ at a point x and let's describe the nearby velocity, at infinitesimal distance dx, to define the elementary types of motion that can be encountered in general. Using Taylor expansion,

$$v(x + dx) = v(x) + \nabla v \cdot dx + O(dx^2), \tag{3.17}$$

the velocity is equal to the velocity at the original point, plus its gradient in the direction of the new point, plus second order terms that will be neglected from now on as we implicitly work in the limit $dx \to 0$. In index notation (3.17) can be rewritten equivalently

$$v_i(x + dx) = v_i(x) + \frac{\partial v_i}{\partial x_j} dx_j, \tag{3.18}$$

where summation on repeated indices (here index j) is implicitly assumed.

The velocity gradient tensor can be divided as the sum of an asymmetric Ω and a symmetric $\mathbb{D}$ tensors

$$\nabla v = \Omega + \mathbb{D}, \quad \frac{\partial v_i}{\partial x_j} = \frac{1}{2}\left(\frac{\partial v_i}{\partial x_j} - \frac{\partial v_j}{\partial x_i}\right) + \frac{1}{2}\left(\frac{\partial v_i}{\partial x_j} + \frac{\partial v_j}{\partial x_i}\right); \tag{3.19}$$

Equation (3.17) can thus be rewritten

$$v(x + dx) = v(x) + \Omega \cdot dx + \mathbb{D} \cdot dx; \tag{3.20}$$

or equivalently from (3.18) in an indexed form.

Let us look at the three terms in (3.20) that sum up to describe the velocity in the neighborhood of a point. The first term describes the *rigid translation* of the small region where all points share the same velocity. The second term is driven by $\boldsymbol{\Omega}$ that is a 3×3 asymmetric tensor, which in Cartesian coordinates reads

$$\boldsymbol{\Omega} = \begin{bmatrix} 0 & +\frac{1}{2}\left(\frac{\partial v_x}{\partial y} - \frac{\partial v_y}{\partial x}\right) & +\frac{1}{2}\left(\frac{\partial v_x}{\partial z} - \frac{\partial v_z}{\partial x}\right) \\ -\frac{1}{2}\left(\frac{\partial v_x}{\partial y} - \frac{\partial v_y}{\partial x}\right) & 0 & +\frac{1}{2}\left(\frac{\partial v_y}{\partial z} - \frac{\partial v_z}{\partial y}\right) \\ -\frac{1}{2}\left(\frac{\partial v_x}{\partial z} - \frac{\partial v_z}{\partial x}\right) & -\frac{1}{2}\left(\frac{\partial v_y}{\partial z} - \frac{\partial v_z}{\partial y}\right) & 0 \end{bmatrix}$$

$$= \frac{1}{2} \begin{bmatrix} 0 & -\omega_z & +\omega_y \\ \omega_z & 0 & -\omega_x \\ -\omega_y & +\omega_x & 0 \end{bmatrix}.$$

Being asymmetric, this tensor is described by 3 independent terms only, and these 3 terms are equal to the components of the vorticity (3.6), module a ½ factor. It was seen before that vorticity was equal to twice the angular velocity, this can be immediately verified noticing that the scalar product $\boldsymbol{\Omega} \cdot d\boldsymbol{x} = \frac{1}{2}\boldsymbol{\omega} \times d\boldsymbol{x}$. Rewriting (3.20) this way

$$\boldsymbol{v}(\boldsymbol{x} + d\boldsymbol{x}) = \boldsymbol{v}(\boldsymbol{x}) + \frac{1}{2}\boldsymbol{\omega} \times d\boldsymbol{x} + \mathbb{D} \cdot d\boldsymbol{x};$$

it is immediate to notice that the second term is an expression that corresponds to a *rigid rotation* with angular velocity $\frac{1}{2}\boldsymbol{\omega}$. Rigid translation and rigid rotation do not produce local deformations; it follows that the relative displacement of the fluid element is only due to the last term. In fact, the symmetric tensor $\mathbb{D}$ is the *rate of deformation tensor*, which in Cartesian coordinates reads

$$\mathbb{D} = \begin{bmatrix} \frac{\partial v_x}{\partial x} & \frac{1}{2}\left(\frac{\partial v_x}{\partial y} + \frac{\partial v_y}{\partial x}\right) & \frac{1}{2}\left(\frac{\partial v_x}{\partial z} + \frac{\partial v_z}{\partial x}\right) \\ \frac{1}{2}\left(\frac{\partial v_x}{\partial y} + \frac{\partial v_y}{\partial x}\right) & \frac{\partial v_y}{\partial y} & \frac{1}{2}\left(\frac{\partial v_y}{\partial z} + \frac{\partial v_z}{\partial y}\right) \\ \frac{1}{2}\left(\frac{\partial v_x}{\partial z} + \frac{\partial v_z}{\partial x}\right) & \frac{1}{2}\left(\frac{\partial v_y}{\partial z} + \frac{\partial v_z}{\partial y}\right) & \frac{\partial v_z}{\partial z} \end{bmatrix}. \qquad (3.21)$$

The scalar product $\mathbb{D} \cdot d\boldsymbol{x}$ represents the (rate of) deformation of the fluid element. The diagonal terms of the tensor are associated with elongation/shortening in the corresponding direction and the off-diagonal are shear motion. The change of volume of the fluid element is due to the combination of elongations/shortening, which is given by the trace of the rate of deformation tensor (the sum of the elements on the diagonal), while a tensor with zero trace does not give a change of volume. The trace of the deformation tensor is the divergence of the velocity field; therefore, it is useful to rewrite (3.20) in its final form as

$$v(x+dx) = v(x) + \frac{1}{2}\omega \times dx + \frac{\nabla \cdot v}{3}\mathbb{I} \cdot dx + \left(\mathbb{D} - \frac{\nabla \cdot v}{3}\mathbb{I}\right) \cdot dx. \qquad (3.22)$$

Expression (3.22), which is also known as the Cauchy-Stokes decomposition, allows recognizing the different elementary movements that combine to describe the motion of an infinitesimal fluid element, which are also sketched in Fig. 3.3. We have already seen that the first term describes rigid translation and the second term is rigid rotation. The third term is pure expansion/compression that is responsible for the local change of volume, a simple scaling effect that does not alter the shape; it will be shown shortly that this term is zero in an incompressible fluid. The last term is the pure deformation, characterized by the relative motion of nearby elements producing a change in shape with no increase or reduction of volume; this term is the only responsible for internal viscous stresses that are due to the relative sliding motion of fluid particles.

Before concluding this section about the description of fluid motion, let us define and specify differences between trajectories and streamlines to ensure using a proper nomenclature. *Trajectories*, as by the normal language, are the curves in space occupied by the same particle during its motion; therefore, trajectories are curves traveled by particles during the time, each trajectory is described by the coordinate $X(t, X_0)$, during time t, of a particle that at an initial time was at X_0. Differently, *streamlines* are curves drawn at one instant of time that are everywhere tangent to the local velocity;

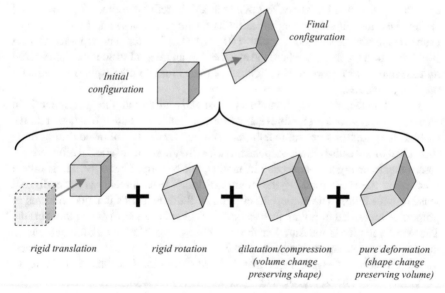

Final configuration

Initial configuration

rigid translation   rigid rotation   dilatation/compression   pure deformation
                                         (volume change   (shape change
                                         preserving shape)   preserving volume)

Fig. 3.3 The act of motion of an infinitesimal volume of fluid from the initial to the final configurations can be subdivided as the combination of four elementary motions

therefore, the streamlines' pattern can vary during the time. In steady flow, trajectories and streamlines coincide and can be used interchangeably. In general, trajectories describe the movement during the time of material particles (what is called Lagrangian description), while streamlines describe the instantaneous flow paths at fixed points in space (Eulerian description); these descriptions are conceptually equivalent although they provide different information.

3.4   Lagrangian and Eulerian Description

The laws of physics are commonly expressed in terms of the conservation of quantities belonging to material elements. It is well known how elementary mechanics deals with individual particles with a given mass that are followed in time while they change their velocity and other properties like, for example, their temperature. Mechanics of rigid bodies also considers the translation and rotation of a given volume made of material elements. This approach is used, in general, for the analysis of solid deformable bodies where the changes in the position, and relative position, of individual material elements are followed during their motion.

Such a natural description of dynamics is thus characterized by the concept of *following the individual elements* that make up the volume of material under analysis. In such a perspective, that is called *Lagrangian description*, each individual material element is identified by its position X_0 at a certain reference time, say $t = 0$, and is then described at subsequent times by its position $X(t, X_0)$ and by the value of properties associated to that element $G(t, X_0)$. The Lagrangian approach is well suited for solid mechanics, where the material has an internal structure characterized by the relative arrangement of individual elements that do not displace excessively one from the other.

In fluid mechanics the situation is drastically different, the individual fluid elements do not have a preferable relative arrangement; they undergo to large relative motion, they mix or separate indefinitely and can hardly be followed during time. Individual blood cells follow independent paths; they separate in arterial bifurcations, some enter in an organ, others enter in another, and so forth. Therefore, a Lagrangian description based on tracking the motion of individual elements is generally not feasible with fluids. The natural description of fluid dynamics is made in terms of properties measured at points fixed in space, which is called *Eulerian description*. The wind velocity is measured at the anemometer position, water temperature is measured at the thermometer position, blood velocity is measured across a valve; all these are Eulerian measurements made at fixed spatial locations that do not refer to the original position of individual particles that pass through.

Indicate with lowercase letters the Eulerian properties measured at time t at a (fixed) spatial location x, the property $g(t, x)$ represents the Eulerian counterpart of the Lagrangian property $G(t, X_0)$. However, both correspond to different descriptions of the same physical property. The Eulerian is more appropriate for measuring and describing the fluid properties. However, conservation laws deal with material elements, and they are more naturally expressed in Lagrangian terms. It is therefore necessary to identify relationships able to transform Lagrangian conservation properties in the Eulerian description of fluid motion.

The relationship between Lagrangian and Eulerian description at the level of individual particles is

$$G(t, X_0) = g(t, X(t, X_0)). \tag{3.23}$$

Relation (3.23) simply states that the properties of the particle X_0 at time t is the same found at the spatial position $X(t, X_0)$ occupied by the particle at time t. Equation (3.23) is important because it provides a bridge between Lagrangian to Eulerian descriptions. Conservation laws are commonly expressed in terms of the time variation of particle properties; for example, acceleration of a fluid particle is the time derivative of velocity of a particle. Relation (3.23) permits to evaluate the time derivative associated with fluid particles in terms of Eulerian quantities. To make it more explicit, remind that Eqs. (3.23) present on the right-hand side a dependence on the vector position X, which corresponds to a dependence on the individual components, e.g., in Cartesian coordinates

$$G(t, X_0) = g(t, X(t, X_0), Y(t, X_0), Z(t, X_0)).$$

Take the time derivative using the chain rule

$$\begin{aligned}
\frac{dG}{dt} &= \frac{\partial g}{\partial t} + \frac{\partial g}{\partial x}\frac{dX}{dt} + \frac{\partial g}{\partial y}\frac{dY}{dt} + \frac{\partial g}{\partial z}\frac{dZ}{dt} \\
&= \frac{\partial g}{\partial t} + v_x\frac{\partial g}{\partial x} + v_y\frac{\partial g}{\partial y} + v_z\frac{\partial g}{\partial z},
\end{aligned} \tag{3.24}$$

where we used the fact that the time derivative of the position is the velocity

$$\frac{dX}{dt} = v.$$

Equation (3.24) can be rewritten in vector form

$$\frac{dG}{dt} = \frac{\partial g}{\partial t} + v \cdot \nabla g. \tag{3.25}$$

The left-hand side of (3.25) is the time derivative in a Lagrangian description, thus following a fluid particle. The right-hand side is the same Lagrangian time derivative written in Eulerian terms, which is sometimes called *material* or *substantial* time derivative. Equation (3.25) states that the material property of a particle passing through a fixed location x can increase either because the property is increasing at that location or because the particle is moving in the direction along which the property increases in space, i.e. when its gradient is aligned with the velocity vector.

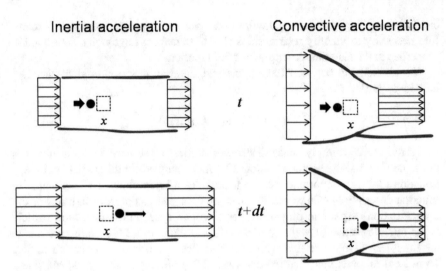

Fig. 3.4 Left: inertial acceleration occurs when velocity increases in time at the fixed spatial location and it can be present even in a straight vessel. Right: convective acceleration occurs when a particle moves towards a region with higher velocity, and it can be present even in steady flows

As a fundamental application, let us apply the (3.25) to the fluid velocity to compute the acceleration of a fluid particle at position x

$$a(t, x) = \frac{\partial v}{\partial t} + v \cdot \nabla v. \tag{3.26}$$

This shows that a particle can accelerate either when velocity increases in time at the position x or, even in steady flow, when the particle is moving toward a region with a higher velocity. This point is sketched in Fig. 3.4, the first term of (3.26) is the *inertial acceleration*, because it is associated with the increase of fluid inertia, the second term is the *convective acceleration*, because it is due to the convection of fluid.

This concept can be extended from individual particles to the integral expressions applied to a finite volume. Integral conservation laws typically apply to a (Lagrangian) material volume of fluid that deforms during its motion, whereas fluid balances are necessarily applied to (Eulerian) spatially defined regions, like a portion of a duct between two cross-sections.

The Reynolds' *Transport theorem* permits to express the time variation of a property associated with a material fluid volume in terms of variations in a spatially fixed volume. Consider a volume of fluid $V_F(t)$ and a fixed volume V that corresponds to the location of the volume of fluid at time t, $V = V_F(t)$, bounded by a fixed surface S. We can prove that

$$\frac{d}{dt} \int_{V_F(t)} G(t) = \int_V \frac{\partial g}{\partial t} dV + \int_S g\mathbf{v} \cdot \mathbf{n} dS, \tag{3.27}$$

where $\mathbf{n}$ is the outward normal to the surface S.

A simple demonstration of (3.27) is as follows. Express the time derivative at the incremental ratio (dt is infinitesimal and implicitly includes the limit to $dt \to 0$)

$$\frac{d}{dt} \int_{V_F(t)} G(t) = \frac{1}{dt} \left[\int_{V_F(t+dt)} g(t+dt) dV - \int_V g(t) dV \right] =,$$

where we can use Eulerian descriptions on the right side because the volumes are instantaneously fixed. Divide the first integral on the right-hand side into two parts

$$= \frac{1}{dt} \left[\int_V g(t+dt) dV + \int_{V_F(t+dt)-V} g(t+dt) dV - \int_V g(t) dV \right] =,$$

the second integral is over the thin space between the volume at time t and that at time $t + dt$, whose infinitesimal volume portion dV is spanned by the infinitesimal surface dS on the boundary of volume V, multiplied by the length travelled normally to that surface in the period dt. In formulas $dV = (\mathbf{v} \cdot \mathbf{n}) dt dS$; thus, the previous formula becomes

$$= \int_V \frac{\partial g}{\partial t} dV + \int_S g(\mathbf{v} \cdot \mathbf{n}) dS,$$

where the first integral combined the formerly first and last terms, and the higher order infinitesimal terms disappeared in the limit of $dt \to 0$. This completed the proof of (3.27).

The transport theorem (3.27) can be rewritten entirely in terms of volume integrals with the aid of the Gauss theorem (3.8),

$$\frac{d}{dt} \int_{V_F(t)} G(t) = \int_V \left[\frac{\partial g}{\partial t} + \nabla \cdot (g\mathbf{v}) \right] dV. \tag{3.28}$$

Equations (3.27) and (3.28) will be fundamental to express the (Lagrangian) conservation laws in terms of (Eulerian) fluid volumes fixed in space.

Part II
Fluid Dynamics: Conservation Laws

Chapter 4
Conservation of Mass

Abstract Conservation of mass represents the first and most fundamental conservation law. It is here expressed in integral terms first, for a generic fluid volume and its application to a cardiac chamber will be presented as a practical example. The integral law is then written in a special form dedicated to vessels when fluid motion is assumed to develop mainly along the vessel direction. Finally, the differential formulation of conservation is obtained. This form applies to every point inside a fluid and enforces a constraint to the velocity vector field.

4.1 Mass Balance in Integral Form

The first law of conservation in order of importance is the conservation of mass. Given a generic material volume of fluid $V_F(t)$, the mass of that volume is given by definition (1.2) that in this case reads

$$\int_{V_F(t)} \rho \, dV,$$

where ρ is the density. Conservation of mass states that the mass of a volume of material cannot vary during time. In formulas it is

$$\frac{d}{dt} \int_{V_F(t)} \rho \, dV = 0. \tag{4.1}$$

The law (4.1) applies to a material volume of fluid (or any continuum) deforming during its motion. Application of the transport theorem (3.27) to (4.1) gives the *integral law of conservation of mass*

$$\int_V \frac{\partial \rho}{\partial t} dV + \int_S \rho \boldsymbol{v} \cdot \boldsymbol{n} dS = 0, \tag{4.2}$$

© The Author(s), under exclusive license to Springer Nature Switzerland AG 2022
G. Pedrizzetti, *Fluid Mechanics for Cardiovascular Engineering*,
https://doi.org/10.1007/978-3-030-85943-5_4

where V is a spatial volume fixed in space and S is the surface surrounding that volume. Equations expressing mass conservation are also called *continuity equation*, because mass conservation ensures the continuity to the material that cannot disappear.

As mentioned before, we will only deal with fluids whose density is constant in time and uniform in space. These fluids are generically referred to here as incompressible fluids (although in rigorous terms, fluids can be incompressible even with spatially variable density). For them, conservation of mass (4.2) simplifies in

$$\int_S v \cdot n \, dS = 0, \tag{4.3}$$

where we remind that the surface S represents the boundary of a fixed region of space.

The integral equation of mass conservation for incompressible fluids, Eq. (4.3) states that given a spatially fixed volume, the amount of fluid that enters through a region on the boundary of such volume must be equal to the amount that leaves through the remaining boundary. For example, in a pipe with perfectly rigid walls, the amount of fluid that enters from the inlet is identical to that exiting at the outlet.

This concept can be expressed in different integral terms for a more immediate application to cavities with varying volumes like, for example, a cardiac chamber. Consider a container of volume $V(t)$ bounded by a surface $S(t)$; this surface can be instantaneously subdivided in a part made by a solid boundary S_b, moving with boundary velocity v_b, and the open sections of area S_{open} allowing the fluid to flow with the fluid velocity v. Being $S = S_b + S_{\text{open}}$, the instantaneous balance (4.3) can be restated as

$$\int_{S_b} v_b \cdot n \, dS + \int_{S_{\text{open}}} v \cdot n \, dS = 0.$$

Consider now that the open boundary of the varying volume $V(t)$ also moves with a velocity v_b that represents its geometric displacement; thus, we can divide the fluid velocity therein as the sum of the boundary velocity plus the relative velocity $v = v_b + (v - v_b)$. The previous balance can be recast as

$$\int_S v_b \cdot n \, dS + \int_{S_{\text{open}}} (v - v_b) \cdot n \, dS = 0,$$

where the first integral is over the entire surface S bounding the volume V. The first term is simply the time variation of the volume and the balance can be written as

$$\frac{dV}{dt} = - \int_{S_{\text{open}}} (\boldsymbol{v} - \boldsymbol{v}_b) \cdot \boldsymbol{n} dS = 0; \tag{4.4}$$

stating that the change of the chamber's volume is given by the net balance of fluid entering and exiting that volume. The minus sign on the right derives from the fact that the normal $\boldsymbol{n}$ points outward. The same equation can be stated in a useful synthetic form as

$$\frac{dV}{dt} = Q_{\text{in}} - Q_{\text{out}}, \quad \begin{cases} Q_{\text{in}} = - \int_{S_{\text{in}}} (\boldsymbol{v} - \boldsymbol{v}_b) \cdot \boldsymbol{n} dS, \\ Q_{\text{out}} = + \int_{S_{\text{out}}} (\boldsymbol{v} - \boldsymbol{v}_b) \cdot \boldsymbol{n} dS; \end{cases} \tag{4.5}$$

where Q_{in} is the total entering discharge across the open section S_{in} through which the flow enters and Q_{out} is the total exiting discharge across S_{out}, $(S_{\text{open}} = S_{\text{in}} + S_{\text{out}})$. It is important to remark that when writing (4.4) or (4.5) the fluid velocities $\boldsymbol{v}$ are Eulerian quantities: values measured relative to a spatially fixed reference. Thus, the discharges (4.5) are written using the relative velocity of fluid with respect to moving boundaries.

As an instructive example, consider a left ventricle (LV), whose total volume is $V_{\text{LV}}(t)$ increases during filling (diastole) while blood enters through the mitral valve of area A_{MV}. Application of (4.4) reads

$$\frac{dV_{\text{LV}}}{dt} = -A_{\text{MV}}(v_{\text{MV}} - v_{b\text{MV}}); \tag{4.6}$$

where v_{MV} is the *fluid* velocity across the mitral valve (normally negative because directed downward, entering the chamber); this is the velocity measured by the imaging methods, like Doppler echocardiography or Phase-Contrast CMR (Cardiac Magnetic Resonance). The value $v_{b\text{MV}}$ is the velocity of the mitral valve boundary (normally positive upward), typically moving the opposite to direction of flow during ventricular expansion thus giving a relative velocity higher than the fluid velocity. In common applications, the latter term is neglected because it is assumed to be much smaller than the fluid velocity. This is commonly realistic, although introduces an approximation that may not be always valid. A balance like (4.6) can be applied to ventricular contraction during flow ejection through the aortic valve, as well as to other chambers or portions of a vessel. It is useful to properly relate measurements of fluid velocity, tissue velocity, and chamber dimension that are linked together by the principle of mass conservation.

4.2 Mass Balance for a Vessel

Consider the flow in a duct where the transversal size is much smaller than the longitudinal extension. The type of fluid motion, in this case, is predominantly one-dimensional (1D), characterized by a velocity component along the vessel direction much larger than the transversal components. In such conditions, it is often useful to consider properties that vary with the position along the vessel and are represent a global behavior for the entire cross-section at that position. Such properties, like area, average velocity, discharge, average pressure, etc., are then expressed as a function of the single spatial coordinate, say x, that defines the position along the vessel and time.

Consider an infinitesimal length dx of such a 1D stream of cross-sectional area $A(t, x)$ and discharge $Q(t, x) = A(t, x)U(t, x)$ being $U(t, x)$ the velocity averaged over the cross-section defined by

$$U(t, x) = \frac{1}{A} \int_A v_x dA. \tag{4.7}$$

Apply the conservation of mass to the quasi-cylindrical short element with volume $V(t, x) = A(t, x)dx$. The flow entering from the first section is $Q_{in} = Q(t, x)$ and that existing is $Q_{out} = Q(t, x) + \frac{\partial Q}{\partial x}dx$, Eq. (4.5) reads

$$\frac{\partial A}{\partial t}dx = Q(t, x) - \left(Q(t, x) + \frac{\partial Q}{\partial x}dx \right),$$

which becomes

$$\frac{\partial A}{\partial t} + \frac{\partial Q}{\partial x} = 0. \tag{4.8}$$

Equation (4.8) expresses the *law of conservation of mass for 1D streams*. It can also be rewritten as

$$\frac{\partial A}{\partial t} + U\frac{\partial A}{\partial x} + A\frac{\partial U}{\partial x} = 0. \tag{4.9}$$

Both Eq. (4.8) and (4.9) express the conservation of mass along a 1D vessel in the absence of lateral inflow/outflow; it states that the flow rate decreases downstream when the vessel area increases in time, and vice versa, as sketched in Fig. 4.1.

In perfectly rigid ducts conservation of mass says the discharge is constant along the vessel. Therefore, conservation of mass between the two arbitrary sections,

Fig. 4.1 Mass conservation implies that flow rate Q does not change along a rigid vessel, while it reduces downstream when the vessel expands

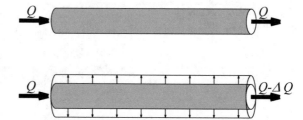

say 1 and 2, of a rigid duct with varying cross-sections permits to evaluate the corresponding changes in the velocity

$$Q_1 = Q_2 \Rightarrow U_1 A_1 = U_2 A_2 \Rightarrow U_2 = U_1 \tfrac{A_1}{A_2} ;$$

stating that velocity increases when the area decreases downstream and vice versa.

In elastic vessels, the increase of area is a consequence of an increase in pressure. Therefore, Eq. (4.8) says that the fluid rate of blood reduces downstream when the generation flow is accompanied by a pressure increase. This is what happens, for example, along the aorta. At the entrance of the aorta, blood enters as an impulse during ventricular systole, this impulse is accompanied by a similar one in terms of pressure (systolic pressure); at the end of the systolic wave, the aortic valve closes and there is no flow during diastole when pressure decreases (diastolic pressure). Therefore, during the systolic impulse the aorta enlarges and accommodates part of the incoming fluid; afterward, pressure decreases, the stored blood is released and the flow rate increases downstream. The result of this phenomenon is the transformation of the sharp flow pulsation at the aortic entrance into a smoother time profile downstream with reduced peak and non-zero flow even during diastole.

4.3 Mass Balance in Differential Form

Equation (4.3) states a balance of the flow rate across the surface bounding a volume V. The same balance can be transformed, with the aid of the Gauss theorem (3.8) in a volume integral

$$\int_V \nabla \cdot v \, dV = 0.$$

Mass conservation applies to any arbitrary volume either large or infinitesimal; if this integral must be zero for any arbitrary volume V then the integrand must be identically zero. This leads to the *law of conservation of mass in differential form*

$$\nabla \cdot v = 0; \tag{4.10}$$

Fig. 4.2 Balance of mass in
an infinitesimal cube

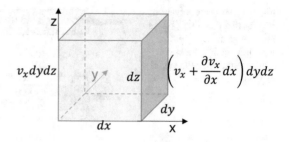

which is commonly called the *continuity equation*. Equation (4.10) implies that the
velocity field of an incompressible flow has zero divergence at every point (velocity
field is solenoidal).

The same result could be obtained by applying (4.3) directly to an infinitesimal
cube. Figure 4.2 shows the balance of flow across the two faces with normal x.
Performing the same operation on the 6 faces

$$-v_x dydz + \left(v_x + \frac{\partial v_x}{\partial x} dx\right) dydz - v_y dxdz + \left(v_y + \frac{\partial v_y}{\partial y} dy\right) dxdz$$
$$-v_z dxdy + \left(v_z + \frac{\partial v_z}{\partial z} dz\right) dxdy = \left(\frac{\partial v_x}{\partial x} + \frac{\partial v_y}{\partial y} + \frac{\partial v_z}{\partial z}\right) dxdydz = 0$$

ends up with the same result (4.10) in Cartesian coordinates

$$\frac{\partial v_x}{\partial x} + \frac{\partial v_y}{\partial y} + \frac{\partial v_z}{\partial z} = 0. \tag{4.11}$$

Condition (4.10), or (4.11), is an important constraint to the possible realization of
the velocity vector field. Looking at the description of flow kinematics in Eq. (3.22)
the velocity field locally can only translate and rotate rigidly and can deform without
change of volume because the flow has zero divergence. If the velocity field converges
from one direction to a point it must similarly diverge in another direction to ensure
that the entire divergence is zero. As a simple intuitive example, if a jet is directed
toward a wall, velocity presents a convergence in that direction because it is positive upstream and zero at the wall for impermeability. As a consequence of mass
conservation, the flow must diverge in the opposite direction, i.e., parallel to the wall
velocity must be directed away from the impact region to create a splash effect on
the wall.

Chapter 5
Conservation of Momentum

Abstract Conservation of momentum represents the fundamental principle of dynamics (the second Newton law); it is the law governing fluid motion, under the constraint of mass conservation. This law is first formulated in integral terms, for a generic fluid volume, which allows to include the effect of momentum on the forces acting on the surrounding boundaries. It is then written in the form dedicated to one-dimensional streams, which regulate the fluid dynamics in vessels presenting the mainly unidirectional motion. The general differential formulation evidences the need to define a constitutive law that specifies the behavior of the material, how internal stresses develop in reaction to fluid motion. The Navier–Stokes equation governing the fluid dynamics of Newtonian fluid is then derived and discussed.

5.1 Momentum Balance in Integral Form

The second law of conservation to consider is the conservation of momentum. This corresponds to the Newton's second law (expressed by $F = ma$ for a single particle) that has to be rewritten for a fluid continuum. Given a generic volume of fluid V_F, the momentum of that volume is defined as

$$\int_{V_F(t)} \rho v \, dV.$$

Conservation of momentum states that the momentum of a material volume can change in time due to the consequence of the application of forces

$$\frac{d}{dt} \int_{V_F(t)} \rho v \, dV = \int_V f \, dV + \int_S \tau \, dS. \tag{5.1}$$

The term on the left-hand side is the variation of momentum (the equivalent of the product between mass and acceleration for a particle). The first term on the right side is the volumetric force that acts at time t on all elements in the volume of fluid,

© The Author(s), under exclusive license to Springer Nature Switzerland AG 2022
G. Pedrizzetti, *Fluid Mechanics for Cardiovascular Engineering*,
https://doi.org/10.1007/978-3-030-85943-5_5

$V = V_F(t)$, and the field $f(x, t)$ is the force per unit volume; the second term is the surface forces applied on the boundary S of the same volume, where τ is the generic stress vector (force vector per unit area). Equation (5.1) is the generalization of the static balance (2.2), which now includes the momentum and the surface force includes other dynamics contributions in addition to pressure.

Application of the transport theorem (3.27) to (5.1) gives

$$\int_V \frac{\partial \rho v}{\partial t} dV + \int_S \rho v(v \cdot n) dS = \int_V f dV + \int_S \tau dS, \tag{5.2}$$

which is the *integral law of conservation of momentum.*

Before moving forward with applications, it can be useful to show that this law can be rewritten in an alternate expression where the first term takes the form of a surface integral. This is feasible in the case of incompressible flows, when the velocity has zero divergence and density is a constant. In this case, application of the Gauss theorem, in particular of its expression (3.15), permits to transform one component of the first integrand in (5.2) into a surface integral

$$\int_V \frac{\partial v_i}{\partial t} dV = \int_V \nabla \cdot \left(x_i \frac{\partial v}{\partial t} \right) dV = \int_S x_i \left(\frac{\partial v}{\partial t} \cdot n \right) dS.$$

Then, insertion of this into (5.2) leads to an expression for the integral law of conservation of momentum (5.2) for incompressible flows

$$\int_S \rho x \left(\frac{\partial v}{\partial t} \cdot n \right) dS + \int_S \rho v(v \cdot n) dS = \int_V f dV + \int_S \tau dS \tag{5.3}$$

that expresses the entire change of momentum in terms of velocity values evaluated on boundaries without the need of knowing velocities in the interior of the volume (Pedrizzetti, 2019). Formulation (5.3) can be useful in several situations involving unsteady flows, as is often the case in cardiovascular circulation.

Symbolically, Eqs. (5.2) or (5.3) is often expressed as

$$\mathbf{I} + \mathbf{M} = \mathbf{G} + \mathbf{\Pi}, \tag{5.4}$$

where the four terms are defined by

$$\begin{aligned} \mathbf{I} &= \int_V \frac{\partial \rho v}{\partial t} dV = \int_S \rho x \left(\frac{\partial v}{\partial t} \cdot n \right) dS, \quad \mathbf{M} = \int_S \rho v(v \cdot n) dS \\ \mathbf{G} &= \int_V f dV \qquad\qquad\qquad\qquad\qquad \mathbf{\Pi} = \int_S \tau dS. \end{aligned} \tag{5.5}$$

The first term is called the local inertia and it can be expressed in terms of volume integral or surface integral, according to Eqs. (5.2) or (5.3), respectively. The second term is the flux of momentum across the boundary, the third term is the volume force, and the last term is the surface force. The balance (5.4) is useful to compute the dynamic forces acting on the boundaries surrounding a moving fluid. This represents an extension of the calculation of static forces. The static analysis (previously seen in Chap. 2) dealt with the last two terms of volumetric and surface forces; the dynamics analysis includes the first two terms associated with the change of momentum due to the fluid velocity, and extends the surface force with the presence of shear stresses.

Let's see now a few instructive examples to demonstrate applications of the dynamic balance for the calculation of dynamic forces, to explore the physical meaning of the terms in (5.4) and the ways of computing them.

Consider a circular duct with a constant cross-section A, presenting a 90° bent on the horizontal plane as sketched in Fig. 5.1. A steady flow, with velocity U, provokes a thrust on the lateral surface of the duct due to flow deviation; that is then transferred to the boundaries where the curve is attached to the rest of the system. Application of the dynamic balance (5.4) permits to compute the force exerted by the flow on the curved duct. First, $\mathbf{I} = 0$, because flow is steady, velocity is constant in time and its time derivative is zero. The flux of momentum at the entrance is given by

$$\mathbf{M}_x = -\int_A \rho v_x^2 dA,\tag{5.6}$$

where the minus sign in front is due to $\mathbf{v} \cdot \mathbf{n} = -v_x$, because velocity v_x enters the volume while the normal is directed outward. When the flow is a mostly unidirectional stream, like in this case, the flux of momentum is often expressed in a compact form in terms of global quantities introducing a velocity-correction coefficient β embodying the effect of velocity variation over the cross-section

$$\mathbf{M}_x = -\rho\beta U^2 A; \quad \beta = \frac{\int_A v^2 dA}{U^2 A}.\tag{5.7}$$

Fig. 5.1 Force on a curved vessel

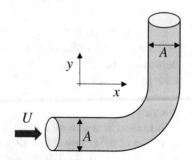

Such a momentum velocity-correction factor β reflects the difference between the average of velocity square and the square of the average velocity. The calculation of the integral of square velocity would require the knowledge of the spatial distribution of velocity that may not be available. In such cases, the introduction of this coefficient allows a simpler formulation based on global properties and transfers the problem of not knowing the transversal profile to a mean of estimating β. This coefficient approaches the unit value when the velocity presents an approximately uniform distribution over the cross-section. Details of the velocity profiles will be studied later; nevertheless, in many situations, the profile does not depart dramatically from the uniform, and (when we have no information to suggest a different number) it can be assumed approximately equal to 1.

Using the same approach, the flux of momentum at the exit is written as

$$M_y = \int_A \rho v_y^2 dA = \rho \beta U^2 A.$$

The volume force, $\mathbf{G}$, assumed due to gravity only, has only the vertical component given by the static weight of the volume. The surface force term is composed of three terms: pressure p_1 acting on the inflow cross-section having the normal to the surface directed in the positive x-direction; pressure p_2 acting on the outflow cross-section with normal directed in the negative y-direction; and the force made by the lateral duct surface that is equal and opposite to the force vector $\mathbf{F} = [F_x, F_y, F_z]$ made by the flow on that surface. In formulas,

$$\Pi_x = p_1 A - F_x, \; \Pi_y = -p_2 A - F_y \; \Pi_z = -F_z.$$

The overall balance (5.3) in the three directions is as follows:

$$-\rho \beta U^2 A = p_1 A - F_x, \; \rho \beta U^2 A = -p_2 A - F_y, \; 0 = -\gamma V - F_z.$$

Therefore, the force made by the flow on the curved vessel is

$$F_x = p_1 A + \rho \beta U^2 A, \; F_y = -\rho \beta U^2 A - p_2 A, \; F_z = -\gamma V. \tag{5.8}$$

A few comments are due to the result (5.8). The force along x is made by the static force $p_1 A$ plus the dynamic force caused by the deviation of the entire incoming momentum, i.e., the impact of the incoming flow onto the bent. The force along y is made by the static force $p_2 A$ that pushes in the negative direction plus the recoil due to the generation of momentum. The vertical force is simply the weight of the fluid volume.

A second instructive example is the case of a rectilinear rigid vessel presenting a reduction of the cross-section along its axis, as shown in Fig. 5.2, from an initial area A_1 to a final $A_2 < A_1$. Let us calculate the terms in the balance (5.3) for this case.

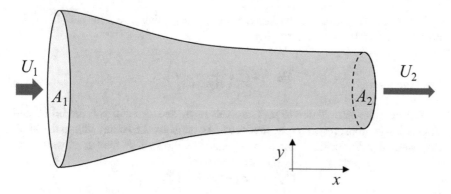

Fig. 5.2 Force on a rectilinear vessel with varying section

The inertial term is non-zero because the flow is unsteady. To this aim, consider the time-varying discharge $Q(t)$ that for mass conservation does not vary along the vessel axis, indicated with x, and can be evaluated at any generic section with area $A(x)$ including inlet or outlet sections

$$Q(t) = \int_{A(x)} v_x dA = U_1(t)A_1 = U_2(t)A_2.$$

The component of the inertial term along the vessel axis becomes

$$I_x = \int_V \frac{\partial \rho v_x}{\partial t} dV = \rho \int_1^2 \int_{A(x)} \frac{\partial v_x}{\partial t} dA dx = \rho \int_1^2 \frac{d}{dt} \left[\int_{A(x)} v_x dA \right] dx.$$

Noticing that the integral reported in brackets in the last term is the discharge $Q(t)$, which does not vary along x. Substitution eventually gives

$$I_x = \rho \frac{dQ}{dt} L, \tag{5.9}$$

where $L = x_2 - x_1$ is the length of the duct under analysis.

The same result (5.9) could be obtained by the second expression in (5.5) based on surface integral

$$I_x = \int_S \rho x \left(\frac{\partial v}{\partial t} \cdot n \right) dS = \int_{A_2} \rho x_2 \frac{\partial v_x}{\partial t} dA - \int_{A_1} \rho x_1 \frac{\partial v_x}{\partial t} dA = \rho(x_2 - x_1) \frac{dQ}{dt}.$$

The flux of momentum is written using (5.7), assuming $\beta = 1$ for simplicity, for both the first and last section to give

$$M_x = \rho Q^2 \left(\frac{1}{A_2} - \frac{1}{A_1} \right).$$

The pressure term is due to pressure values p_1 and p_2 acting on the initial and final cross-sections, respectively, plus the force made by the lateral wall to the fluid, that is equal and opposite to the force F_x made by the fluid on the lateral wall

$$\Pi_x = p_1 A_1 - p_2 A_2 - F_x.$$

Inserting these terms in the balance (5.3) along x, the force made by flow on the vessel is

$$F_x = p_1 A_1 - p_2 A_2 - \rho Q^2 \left(\frac{1}{A_1} - \frac{1}{A_2} \right) - \rho \frac{dQ}{dt} L. \qquad (5.10)$$

The first two terms give the static force acting on the two cross-sections that depend on the values of pressure and not on the flow velocity (although we will see later that the difference of pressure is actually consequent on the properties of the flow). The third term is the next flux of momentum; it is negative and represents the reaction effect due to the higher flux of momentum at the exit than at the inlet (despite the flux of mass is the same). The last term reflects the force associated with the inertia (acceleration/deceleration) of the whole volume of fluid contained inside the vessel.

This example can also be used to evaluate the force when the contraction is very sharp as it can be the case of an orifice placed transversally in the vessel, which can represent a model for a cardiovascular valve. In that case, formula (5.10) provides an estimation for the force pushing on the upstream surface of the orifice wall.

As a further simple example compute the force produced by a fluid jet directed toward a planar surface. With reference to Fig. 5.3, consider a steady jet with average velocity U and cross-area A, directed perpendicular to a flat plate. Define the volume as bounded by the inlet section, the streamlines at the boundary of the jet adjacent to the external environment, assumed as air at atmospheric pressure, the plate, and the exit sections where velocities are directed parallel to the plate. The balance for this volume in the direction of the jet is straightforward to compute. Inertia is zero because the flow is steady, the flux of momentum at the inlet is $M = -\rho \beta U^2 A$, where β can be confidently assumed $\beta \cong 1$ being it a free jet. While the flux at the outlet does not contribute to this direction (and is zero for symmetry in the transversal direction). Body force (gravity) is also zero in this direction. Surface forces are not present on the lateral boundaries because the pressure is zero (equal to the atmospheric value, which is taken as the reference zero value). Similarly, pressure is also zero at the inlet section because the pressure is constant inside the jet (this will be demonstrated rigorously later) that is surrounded by atmospheric pressure. Therefore, surface force

Fig. 5.3 Force of a jet impacting on a flat plate

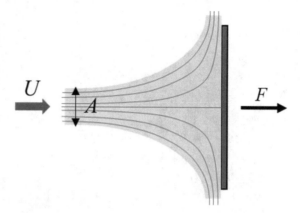

is non-zero only on the obstacle and it is opposite to the force F made by flow on it, $\Pi = -F$.

Insertion of these findings in the balance (5.3) shows that the force on the plate is only given by the deviation of the incoming momentum

$$F = \rho U^2 A = \rho Q U. \tag{5.11}$$

These simple cases were presented to show the appropriate simple means of evaluating the terms in (5.4) under typical conditions, at least for a first approximation. In general, the integrals (5.5) provide the mean for an accurate evaluation of all terms.

The balance (5.2) can be extended to provide the balance of angular momentum. In which case every term must be multiplied with the corresponding arm of the force. The details of this extension are not reported here as they are not of primary interest for the topic of this book and do not bring conceptual challenges. However, such extension is immediate in most situations using the same overall approach described above.

5.2 Momentum Balance for a Vessel

Following the same procedure that we used above for conservation of mass, let us rearrange the balance of momentum (5.2) for the special important case of flow in a vessel where fluid motion develops predominantly along the direction of the duct axis. Indicate again with x the longitudinal direction along the vessel and with $A(t, x)$ the area of the cross-section that can vary along the axis and in time, and the transversal velocities are assumed to be negligible with respect to the longitudinal one whose average value is $U(t, x)$.

Consider an infinitesimal length dx of such a vessel and let us evaluate the component along the vessel of the individual terms in (5.4), recalling that the balance is

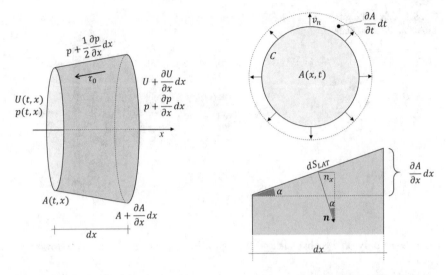

Fig. 5.4 Indication of the quantities involved in the momentum balance for an infinitesimal portion of a vessel

made on a spatial volume (instantaneously fixed). The volume under analysis is sketched in Fig. 5.4; it is bounded upstream by the cross-section of area $A(t, x)$, where velocity is $U(x, t)$ and average pressure is $p(x, t)$. It is bounded downstream by the cross-section of area $A(t, x + dx) = A(t, x) + \frac{\partial A}{\partial x} dx$, where velocity is $U(t, x + dx) = U(t, x) + \frac{\partial U}{\partial x} dx$, and pressure is $p(t, x + dx) = p(t, x) + \frac{\partial p}{\partial x} dx$; it is also bounded laterally by the perimeter curve $C(t, x)$ that extends over the length dx.

The inertial term is integrated over the volume $dV = d A dx$ and reads

$$I = \int_V \frac{\partial \rho v_x}{\partial t} dV = \rho \int_A \frac{\partial v_x}{\partial t} d A dx = \rho \frac{\partial}{\partial t} \int_A v_x d A dx = \rho A \frac{\partial U}{\partial t} dx. \qquad (5.12)$$

The flux of momentum across the two cross-sections and the lateral contour

$$M = - \int_{A(x)} \rho v_x^2 d A + \int_{A(x) + \frac{\partial A}{\partial x} dx} \rho \left(v_x + \frac{\partial v_x}{\partial x} dx \right)^2 d A + \int_{C(x)} \rho v_x v_n d C dx =$$

is simplified assuming that the velocity is uniform over the cross-section, which means $v_x = U$ and $\beta = 1$, and ignoring all terms of order dx^2

$$= -\rho U^2 A + \rho \left(U^2 + 2U\frac{\partial U}{\partial x}dx\right)\left(A + \frac{\partial A}{\partial x}dx\right) + \rho U \int_C v_n dC dx =$$

$$= -\rho U^2 A + \rho U^2 A + \rho 2U\frac{\partial U}{\partial x}Adx + \rho U^2 \frac{\partial A}{\partial x}dx + \rho U \int_C v_n dC dx =$$

The last integral above can be rewritten (see Fig. 5.4) considering that the integral of $v_n dC$ is the rate of increase of the cross-area, $\frac{\partial A}{\partial t}$. Rearranging all terms.

$$= 2\rho U\frac{\partial U}{\partial x}Adx + \rho U^2 \frac{\partial A}{\partial x}dx + \rho U \frac{\partial A}{\partial t}dx =$$

$$= \rho U\frac{\partial U}{\partial x}Adx + \rho U \left(A\frac{\partial U}{\partial x} + U\frac{\partial A}{\partial x} + \frac{\partial A}{\partial t}\right)dx.$$

The term in bracket is equal to zero for mass conservation (4.9) and the whole flux of momentum becomes

$$\mathbf{M} = \rho U \frac{\partial U}{\partial x}Adx. \tag{5.13}$$

Surface forces are composed of pressure acting on the two cross-sections, the wall shear stress acting on the lateral surface, and the contribution of pressure on the lateral surface, which may present a longitudinal component when the cross-section is not constant. Surface forces are

$$\Pi = pA - \left(p + \frac{\partial p}{\partial x}dx\right)\left(A + \frac{\partial A}{\partial x}dx\right) - \tau_0 C dx + \left(p + \frac{1}{2}\frac{\partial p}{\partial x}dx\right)n_x dS_{LAT} =$$

Last but one term has is the average wall shear stress, $-\tau_0$, exerted by the lateral solid boundary to the fluid; it has a negative sign as τ_0 conventionally indicates the stress made by fluid to the wall. Last term is the x-component of the pressure force on the lateral surface (pressure value is taken as the mean between x and $x + dx$) and n_x is the x component of the normal unit vector on the lateral surface dS_{LAT} (notice that here we did not use the simplification $dS_{LAT} = Cdx$ to remark the relevance of the change of cross-section along dx, the tilting of the later surface, whereas C is the average value along the length dx). It can be noticed (see Fig. 5.4) that the product $n_x dS_{LAT}$ corresponds to the projection of the lateral surface on the cross plane, thus it is equal to the change of cross-surface $n_x dS_{LAT} = \frac{\partial A}{\partial x}dx$.

Ignoring the higher order terms in dx and simplifying

$$\Pi = pA - pA - \frac{\partial p}{\partial x}Adx - p\frac{\partial A}{\partial x}dx - \tau_0 C dx + p\frac{\partial A}{\partial x}dx = -A\frac{\partial p}{\partial x}dx - \tau_0 C dx. \tag{5.14}$$

Finally, the volume force, assumed imputable to gravity only $f = -\gamma\nabla z$, is obtained by the component along the vessel $f_x = -\gamma\frac{\partial z}{\partial x}$ integrated over the volume and gives (ignoring terms containing dx^2)

$$G = -\gamma A \frac{\partial z}{\partial x} dx, \tag{5.15}$$

where z stands for the vertical direction aligned with gravity. Combine all terms (5.12)–(5.15) of the momentum balance (5.4) and divide by $\rho A dx$ to obtain

$$\frac{\partial U}{\partial t} + U \frac{\partial U}{\partial x} = -\frac{\partial}{\partial x}\left(\frac{p}{\rho} + gz\right) - \frac{\tau_0}{\rho}\frac{C}{A}. \tag{5.16}$$

Equation (5.16) is the *law of conservation of momentum for 1D streams*, under the assumption of uniform velocity over the cross-section and in presence of gravity only.

The left-hand terms represent the (Lagrangian) acceleration of a 1D fluid element moving with velocity U expressed in terms of (Eulerian) derivatives in a fixed frame of reference. The first term on the right-hand side is the driving force that can be due either to a pressure gradient (negative, higher upstream, and lower downstream) or to a difference of quote. This expression underlines again that pressure gradient and gravity play the same role in a fluid motion. It is a common habit to use a generalized pressure that includes gravity. Then (5.16) is usually rewritten without explicit mention of gravity (or another conservative force) as

$$\frac{\partial U}{\partial t} + U \frac{\partial U}{\partial x} = -\frac{1}{\rho}\frac{\partial p}{\partial x} - \frac{\tau_0}{\rho}\frac{C}{A}, \tag{5.17}$$

where p stands for the generalized pressure equal to $p + \gamma z = \gamma h$, where h is the static head previously introduced with Eq. (2.6) in fluid statics. Then, if needed, the actual pressure can be recovered simply by removing the static contribution due to gravity.

The last term is the friction on the lateral walls. This term depends on the velocity profile near the wall, as shown for example by Eq. (1.9) for a Newtonian fluid. The 1D model, however, deals with the mean velocity only, which is assumed to be uniform over the cross-section and does not provide information about transversal velocity gradient. Therefore, the friction terms must be provided by a model based on the hypothesis on the velocity profile near the wall and the property of the fluid. In some theoretical studies, it is often neglected to evaluate the behavior in the limit case of flow without viscous resistance. In some applications, it is expressed as proportional to the ratio between the velocity field U and the diameter of the vessel with the proportionality coefficient defined by a model based on phenomenological assumptions.

5.3 Momentum Balance in Differential Form for a Continuum: Cauchy Equation

The laws of conservation of momentum presented so far concerned balances over extended regions and were not intended to describe fluid motion in its fine space–time details. This can be achieved when the balance of momentum is formulated at the level of infinitesimal volumes or, more precisely, in differential form.

To this purpose, start from the balance of momentum expressed in global terms (5.2), divided by the constant density for convenience,

$$\int_V \frac{\partial v}{\partial t} dV + \int_S v(v \cdot n) dS = \frac{1}{\rho} \int_V f dV + \frac{1}{\rho} \int_S \tau dS. \qquad (5.18)$$

This balance is made up of four integrals, two on the volume of fluid and two on the surrounding boundary. We want to transform the surface integrals, second and fourth terms in (5.18), as volume integrals, such that all integrals refer to the same arbitrary volume and the equality can be extended from the integral to the integrand terms. This is straightforward for the second integral in (5.18) where the generic ith component can be transformed as follows:

$$\int_S v_i v \cdot n dS = \int_V \nabla \cdot (v_i v) dV = \int_V (v_i \nabla \cdot v + v \cdot \nabla v_i) dV = \int_V v \cdot \nabla v_i dV. \quad (5.19)$$

The first equality used the Gauss theorem (3.8) applied to the vector field $v_i v$; the second equality is immediate to verify using the derivative of a product for vector terms, and last equality follows after canceling the terms $\nabla \cdot v$ that is identically zero for mass conservation (4.10).

The last term in (5.18) contains the stress vector τ acting on the surface dS. Apparently, at a point there are infinite stress vectors that can act on surfaces with different orientations, and the identification of the vector τ that acts on the specific surface dS may look like a complex task. However, such an infiniteness is only apparent because there is a single stress "state" about a point and the value of all these individual vectors comes from a combination of such stress state in relation to the orientation of the surface. Indeed, it can be demonstrated that the stress vector acting on a surface with normal n can be expressed in general as

$$\tau = \mathbb{T} \cdot n, \qquad (5.20)$$

where $\mathbb{T}$ is the stress tensor. It characterizes the stress state at a point, such that the stress vector at that point acting on a surface with normal n is obtained by projecting the stress tensor over the direction n, as by (5.20).

Fig. 5.5 Cauchy tetrahedron

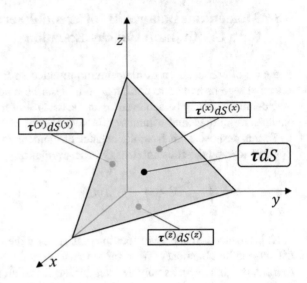

Result (5.20) is usually demonstrated using the Cauchy tetrahedron, which is built by the original surface dS and its projection on the Cartesian planes as shown in Fig. 5.5.

Indicate with $\boldsymbol{\tau}^{(x)}$ the stress vector acting on the surface $dS^{(x)}$, which is the projection of dS on the y-z plane perpendicular to the x-axis; and the same for the other coordinate axes. First, we want to see whether the stress $\boldsymbol{\tau}$ on the original surface can be expressed as a combination of the stresses $\boldsymbol{\tau}^{(x)}$, $\boldsymbol{\tau}^{(y)}$, $\boldsymbol{\tau}^{(z)}$ acting on the surfaces normal to the Cartesian axes. Balance of the forces acting on the tetrahedron gives the equivalence of the surface forces

$$\boldsymbol{\tau}dS = \boldsymbol{\tau}^{(x)}dS^{(x)} + \boldsymbol{\tau}^{(y)}dS^{(y)} + \boldsymbol{\tau}^{(z)}dS^{(z)}. \tag{5.21}$$

It is then easy to verify by simple geometry that the projection $dS^{(i)} = dSn_i$, where the n_i is the ith component of the normal $\boldsymbol{n}$ to the surface dS. Introducing this into (5.21) gives

$$\boldsymbol{\tau} = \boldsymbol{\tau}^{(x)}n_x + \boldsymbol{\tau}^{(y)}n_y + \boldsymbol{\tau}^{(z)}n_z. \tag{5.22}$$

If you define the stress tensor as a tensor made by the three stress vectors placed in column $\tau_i^{(j)} = \mathbb{T}_{ij}$ then Eq. (5.22) corresponds to (5.20) that is thus proven. Using expression (5.20), the fourth term in (5.18) can be rewritten as a volume integral through the Gauss theorem

$$\frac{1}{\rho}\int_S \boldsymbol{\tau}dS = \frac{1}{\rho}\int_S \mathbb{T}\cdot\boldsymbol{n}dS = -\frac{1}{\rho}\int_V \nabla\cdot\mathbb{T}dV, \tag{5.23}$$

where the minus comes out because the normal in Gauss theorem is outward directed while it is common to consider here the inward normal to agree with the convention that a positive pressure makes a force directed toward a surface (inward normal). Introduction of (5.19) and (5.23) in the momentum balance (5.18) allows rewriting in terms of volume integrals

$$\int_V \frac{\partial v}{\partial t} dV + \int_V v \cdot \nabla v dV = \frac{1}{\rho} \int_V f dV - \frac{1}{\rho} \int_V \nabla \cdot \mathbb{T} dV.$$

This equality must be valid for an arbitrary volume, including any infinitesimal volume, and therefore the balance must apply to the integrands as well

$$\frac{\partial v}{\partial t} + v \cdot \nabla v = \frac{1}{\rho} f - \frac{1}{\rho} \nabla \cdot \mathbb{T}. \tag{5.24}$$

Equation (5.24) is the *Cauchy equation* that expresses the law of conservation of momentum for a continuum.

The same result could be obtained in Cartesian coordinates by applying the balance of momentum (5.18) to an infinitesimal cube of volume as shown in Fig. 5.6.

Consider, for example, the x-component (then results can be immediately extended to the other components). The first term in (5.18) applied to the infinitesimal cube, $V = dxdydz$, becomes

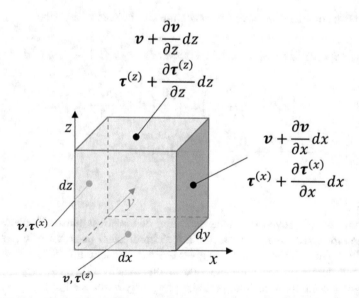

Fig. 5.6 Balance of momentum in an infinitesimal cube (values on the faces perpendicular to y are not shown to for clarity)

$$\int_V \frac{\partial v_x}{\partial t} dV = \frac{\partial v_x}{\partial t} dx dy dz. \tag{5.25}$$

The second term includes the fluxes of momentum on the 6 faces

$$\int_S v_x(\boldsymbol{v} \cdot \boldsymbol{n}) dS$$

$$= -v_x v_x dy dz - v_x v_y dx dz - v_x v_z dx dy + \left(v_x + \frac{\partial v_x}{\partial x} dx\right)\left(v_x + \frac{\partial v_x}{\partial x} dx\right) dy dz$$

$$+ \left(v_x + \frac{\partial v_x}{\partial y} dy\right)\left(v_y + \frac{\partial v_y}{\partial y} dy\right) dx dz + \left(v_x + \frac{\partial v_x}{\partial z} dz\right)\left(v_z + \frac{\partial v_z}{\partial z} dz\right) dx dy$$

$$= \left(v_x \frac{\partial v_x}{\partial x} + v_x \frac{\partial v_x}{\partial x} + v_x \frac{\partial v_y}{\partial y} + v_y \frac{\partial v_x}{\partial y} + v_x \frac{\partial v_z}{\partial z} + v_z \frac{\partial v_x}{\partial z}\right) dx dy dz$$

$$= (\boldsymbol{v} \cdot \nabla v_x + v_x \nabla \cdot \boldsymbol{v}) dx dy dz$$

$$= \boldsymbol{v} \cdot \nabla v_x dx dy dz \tag{5.26}$$

The third term

$$\int_V f_x dV = f_x dx dy dz. \tag{5.27}$$

Last term combines the stress forces on the six surfaces of the cube

$$\int_S \tau_x dS = \tau_x^{(x)} dy dz + \tau_x^{(y)} dx dz + \tau_x^{(z)} dx dy - \left(\tau_x^{(x)} + \frac{\partial \tau_x^{(x)}}{\partial x} dx\right) dy dz +$$

$$- \left(\tau_x^{(y)} + \frac{\partial \tau_x^{(y)}}{\partial y} dy\right) dx dz - \left(\tau_x^{(z)} + \frac{\partial \tau_x^{(z)}}{\partial z} dz\right) dx dy$$

$$= -\left(\frac{\partial \tau_x^{(x)}}{\partial x} + \frac{\partial \tau_x^{(y)}}{\partial y} + \frac{\partial \tau_x^{(z)}}{\partial z}\right) dx dy dz$$

$$= -(\nabla \cdot \mathbb{T})_x dx dy dz \tag{5.28}$$

where here we employed the same definition of the stress tensor $\mathbb{T}$ made up by the three stress vectors relative to the three coordinates. Insertion of expressions (5.25)–(5.28) into the balance (5.18) gives again the Cauchy Eq. (5.24).

The two terms on the left-hand side of the Cauchy Eq. (5.24) represent the Lagrangian acceleration of fluid particles previously introduced in Eq. (3.26). The two terms on the right-hand side are the forces acting on such particles, caused by intrinsic volumetric forces and by the stresses made by the neighboring fluid elements.

The same procedure may now be performed for the conservation of angular momentum by writing a similar expression including the arms of the individual terms. The derivation is somehow lengthy and is not reported in detail here because it does not produce further differential equations; the result is remarkably simple. The *conservation of angular momentum* implies that *the stress tensor $\mathbb{T}$ is a symmetric tensor*. This reduces the complexity of the 3×3 stress tensor from 9 components to 6 independent components.

5.4 Momentum Balance for Newtonian Fluids: Navier–Stokes Equations

Let us recapitulate the set of equations describing the mechanics of a continuum (still we have not used any argument that this continuum is a fluid, only that it is incompressible). This is a system composed of the conservation of mass (4.10) (the continuity equation) and the conservation of momentum (5.24) (Cauchy equation)

$$\begin{cases} \nabla \cdot v = 0, \\ \frac{\partial v}{\partial t} + v \cdot \nabla v = \frac{1}{\rho} f - \frac{1}{\rho} \nabla \cdot \mathbb{T}; \end{cases} \tag{5.29}$$

which is a set of 4 scalar equations. The unknowns are the 3 components of the velocity vector and the 6 components of the stress tensor, resulting in a number of 9 total unknowns. A set of 4 equations with 9 unknowns is not a closed set, it cannot be solved until some additional information is provided.

The set of Eqs. (5.29) is valid for a generic continuum; it applies to both solids and fluids. Before being able to face any physical problem, it is, therefore, necessary to define the material that constitutes such a continuum. We must introduce information about such a material because the outcome depends on whether this behaves like a fluid, a solid, or else. In other terms, we must introduce the *constitutive law* that specifies how internal stresses develop in consequence of the deformation of the material. In the following, we will develop a constitutive law for fluids, following the general properties described in Sect. 1.2.

The first information for specifying the constitutive law comes from the statics of fluids. In Chap. 2, we have seen that under static conditions the stresses acting on a surface is to pressure that acts normally toward the surface; in formula $\tau = p n$ (with the convention of the inward normal). Comparison with (5.20) immediately implies that in the limit case of static conditions, the stress tensor is required to take the form

$$\mathbb{T} = p\mathbb{I} = p \begin{bmatrix} 1 & 0 & 0 \\ 0 & 1 & 0 \\ 0 & 0 & 1 \end{bmatrix};$$

where $\mathbb{I}$ is the identity matrix.

The second information comes from the kinematics of fluids. In Sect. 3.3, we have shown that motion is composed of rigid translation and rotation plus a pure deformation. The latter is the only elementary action that involves the relative motion of fluid elements, thus the only action that can be responsible for friction and internal stresses. Therefore, we can express in general the constitutive law for a fluid as

$$\mathbb{T} = p\mathbb{I} + f(\mathbb{D}), \tag{5.30}$$

where $\mathbb{D}$ is the symmetric deformation rate tensor (3.21) and f is a general functional dependence. Relationship (5.30) states that, once the function f is defined, the 6 unknowns present in the tensor $\mathbb{T}$ can be expressed in terms of derivatives of the velocity field plus a single unknown, pressure p. Thus Eq. (5.30) provides a closure (balance between equations and unknowns) to the system (5.29). Fluids following the law (5.30), where stress forces are given by the rate of deformation, are called Stokes fluids.

The third information comes from the definition of viscosity for a Newtonian fluid. In Sect. 1.2, we showed that the stress due to shear flow along x on a surface with normal y is given by formula (1.9); which that can be restated with the current formalism as

$$\mathbb{T}_{xy} = -\mu \frac{\partial v_x}{\partial y}.$$

This expression is not symmetric and violates the conservation of angular momentum; however, it can easily be made symmetric as

$$\mathbb{T}_{xy} = -\mu \left(\frac{\partial v_x}{\partial y} + \frac{\partial v_y}{\partial x} \right);$$

without contradicting the experimental result (1.9) because the transversal velocity v_y was zero. This is an off-diagonal term of a form compatible with (5.30), suggesting that the function $f(\mathbb{D})$ appearing in (5.30) is a linear one for Newtonian fluids.

Combining this set of information, the *constitutive law for Newtonian fluids* is written in general as

$$\mathbb{T} = p\mathbb{I} - 2\mu\mathbb{D}; \tag{5.31}$$

or in individual Cartesian components

$$\mathbb{T} = \begin{bmatrix} p - 2\mu\frac{\partial v_x}{\partial x} & -\mu\left(\frac{\partial v_x}{\partial y} + \frac{\partial v_y}{\partial x}\right) & -\mu\left(\frac{\partial v_x}{\partial z} + \frac{\partial v_z}{\partial x}\right) \\ -\mu\left(\frac{\partial v_x}{\partial y} + \frac{\partial v_y}{\partial x}\right) & p - 2\mu\frac{\partial v_y}{\partial y} & -\mu\left(\frac{\partial v_y}{\partial z} + \frac{\partial v_z}{\partial y}\right) \\ -\mu\left(\frac{\partial v_x}{\partial z} + \frac{\partial v_z}{\partial x}\right) & -\mu\left(\frac{\partial v_y}{\partial z} + \frac{\partial v_z}{\partial y}\right) & p - 2\mu\frac{\partial v_z}{\partial z} \end{bmatrix}.$$

Let us now look how the stress tensor expression given by the constitutive law (5.31) reflects in the surface force term in the Cauchy Eq. (5.24). The surface force is the divergence of the stress tensor, consider the x component for simplicity

$$\begin{aligned} -\nabla \cdot \mathbb{T}|_x &= -\left(\frac{\partial \mathbb{T}_{xx}}{\partial x} + \frac{\partial \mathbb{T}_{xy}}{\partial y} + \frac{\partial \mathbb{T}_{xz}}{\partial z}\right) \\ &= -\frac{\partial p}{\partial x} + 2\mu\frac{\partial^2 v_x}{\partial x^2} + \mu\frac{\partial^2 v_x}{\partial y^2} + \mu\frac{\partial^2 v_y}{\partial xy} + \mu\frac{\partial^2 v_x}{\partial z^2} + \mu\frac{\partial^2 v_z}{\partial zx} \\ &= -\frac{\partial p}{\partial x} + \mu\left(\frac{\partial^2 v_x}{\partial x^2} + \frac{\partial^2 v_x}{\partial y^2} + \frac{\partial^2 v_x}{\partial z^2}\right) + \mu\frac{\partial}{\partial x}\left(\frac{\partial v_x}{\partial x} + \frac{\partial v_y}{\partial y} + \frac{\partial v_z}{\partial z}\right) \\ &= -\frac{\partial p}{\partial x} + \mu\nabla^2 v_x. \end{aligned} \tag{5.32}$$

Insertion of this result into the Cauchy Eq. (5.24) gives the equations for conservation of momentum for Newtonian fluids: the *Navier–Stokes equation*

$$\frac{\partial \mathbf{v}}{\partial t} + \mathbf{v} \cdot \nabla\mathbf{v} = \frac{1}{\rho}\mathbf{f} - \frac{1}{\rho}\nabla p + \nu\nabla^2\mathbf{v}; \tag{5.33}$$

where $\nu = \frac{\mu}{\rho}$ is the kinematic viscosity previously defined in (1.10). This equation is also called the *law of motion* for an incompressible Newtonian fluid, and represents the expression of the Newton's second law rearranged for a Newtonian fluid. The left-hand side is the acceleration of a fluid particle, the terms on the right-hand side are the forces, per unit mass. They are the volumetric force, the thrust due to pressure difference, and the resistance force due to internal viscous friction, respectively.

As discussed in Chap. 1, blood is a complex material for which the assumption of a Newtonian constitutive relations is approximate. The reliability of this approximation was discussed therein and it is not recalled here. In what follows we will limit our analysis to Newtonian fluids, which are the foundation for understanding most of the flow phenomena in the heart and large blood vessels. It is remarked and must be understood that this is a model for blood flow that may be good in many situations, which is not perfect in general and can be inappropriate in other situations.

When dealing with gravitational volume forces only, as it is common in the circulation, we have seen that these can be rewritten in gradient form $\mathbf{f} = \gamma\nabla z$. Therefore, it is common to formally include the conservative force in the pressure term substituting the quantity $p + \gamma z$ with the sole pressure symbol p that represents the static head and is now meant to include gravity

$$\frac{\partial \mathbf{v}}{\partial t} + \mathbf{v} \cdot \nabla\mathbf{v} = -\frac{1}{\rho}\nabla p + \nu\nabla^2\mathbf{v}. \tag{5.34}$$

The set of equations given by continuity Eq. (4.10) and Navier–Stokes Eq. (5.34) is now a formally complete set with the same number of equations (4 scalar equations) and unknowns (the 3 components of the velocity vector and pressure)

$$
\begin{cases}
\nabla \cdot \boldsymbol{v} = 0, \\
\frac{\partial \boldsymbol{v}}{\partial t} + \boldsymbol{v} \cdot \nabla \boldsymbol{v} = -\frac{1}{\rho}\nabla p + \nu \nabla^2 \boldsymbol{v}.
\end{cases}
\tag{5.35}
$$

This system of equations, continuity, and motion must be completed with the appropriate boundary conditions. The Navier–Stokes equation is a partial differential equation containing second-order derivatives for velocity; therefore, roughly speaking, it requires two boundary conditions for velocity. The first condition is the impermeability at the boundary between the fluid and solid; this means that the normal component of the relative velocity must be zero (normal velocity equal to that of the boundary, when it is moving). The second condition is the adherence to the wall; this means that also the relative tangential velocity must go to zero at the wall. Therefore, the entire relative velocity vector must vanish at a boundary. However, the two conditions are deeply different. Adherence is a purely viscous phenomenon; the appearance of this additional condition is congruent with the fact that it is consequent to the presence of the viscous terms that is the only one containing second-order derivatives.

The viscous, frictional term in the Navier–Stokes equation produces energy dissipation. In a universal perspective, the total energy is conserved and friction is a mechanism of transformation of kinetic energy into heat. Therefore, from the mechanical perspective, friction provokes a dissipation, a reduction of the mechanical energy that is lost because it will be transformed into a different form of energy that is not considered in a balance limited to mechanical aspects.

The kinematic viscosity is a small coefficient (for reference $\nu = 10^{-6}$ m^2/s $= 10^{-2}$ cm^2/s for water). Therefore, especially far from the boundaries, the viscous terms can be sometimes neglected and fluid behaves in most cases approximately like a non-viscous one. Therefore, the limiting case when viscosity is zero, $\nu = 0$, can be useful as a model in numerous applications. In this asymptotic limit, we talk of *ideal fluids* (also called inviscid or frictionless). The equation of motion for ideal fluids is the *Euler equation*

$$
\frac{\partial \boldsymbol{v}}{\partial t} + \boldsymbol{v} \cdot \nabla \boldsymbol{v} = -\frac{1}{\rho}\nabla p,
\tag{5.36}
$$

which differs from the Navier–Stokes Eq. (5.34) for the absence of the viscous term only. The Euler equation does not present friction and therefore conserves mechanical energy. Thus, it describes reversible phenomena: if the velocity field $\boldsymbol{v}(t, \boldsymbol{x})$ is solution of the Euler equation forward in time, then the reversed field $-\boldsymbol{v}(-t, \boldsymbol{x})$ is also a solution backward in time. This was not true for the Navier–Stokes equation due to the friction term that does not allow time reversal (reverse flow also would have friction; it certainly does not transform heat back into kinetic energy).

This apparently small change brings along another important difference between Euler and Navier–Stokes, the fact that the former is a first-order partial differential equation because it contains first-order derivatives only, which reflects into the fact that only one boundary condition can be imposed for the velocity. Namely, only the impermeability condition applies to the Euler equation while adherence does not. This is perfectly physically consistent because the adherence is a viscous phenomenon; ideal flows have no viscosity and cannot have viscous adherence.

The Euler equation is important because it allows some simple solutions to specific applications; however, care must be taken when applying the approximation of ideal flow. It can be used over short paths, where the small viscosity may be effectively negligible, and far from boundaries outside the regions influenced by viscous adherence. This aspect will be taken up several times when dealing with specific applications throughout the book.

A last consideration about Navier–Stokes (and Euler) equation regards the frequent case of flows that are essentially unidirectional, where one velocity component is much larger than the others. To fix the ideas, consider a motion that is predominantly along the x-direction, thus the transversal velocity components are negligible with respect to the streamwise, $v_y \cong 0$ and $v_z \cong 0$. Now, write the Navier–Stokes equation over a direction transversal to the directions of motion, for example, the y direction

$$\frac{\partial v_y}{\partial t} + v_x \frac{\partial v_y}{\partial x} + v_y \frac{\partial v_y}{\partial y} + v_z \frac{\partial v_y}{\partial z} = -\frac{1}{\rho}\frac{\partial p}{\partial y} + \nu\left(\frac{\partial^2 v_y}{\partial x^2} + \frac{\partial^2 v_y}{\partial y^2} + \frac{\partial^2 v_y}{\partial z^2}\right); \quad (5.37)$$

If we can neglect the velocity v_y and its derivatives; in other words, if *streamlines are straight and parallel*, then Eq. (5.37) reduces to

$$\frac{\partial p}{\partial y} = 0, \quad (5.38)$$

which, in presence of gravity, has the meaning

$$\frac{\partial}{\partial y}(p + \gamma z) = 0.$$

This is a general and important result. In regions where fluid motion is straight and parallel, the static head (2.6), given by pressure plus gravity if the latter is present, remains constant transversal to the direction of motion.

In simpler terms, the law of fluid statics (2.5) holds along the directions without motion (transversal to flow). This simple fact was sometimes used in Sect. 5.1 when computing dynamic forces. It also tells that the average pressure value in the equation for a vessel (5.17) is actually constant over the cross-section. This basic concept will be used several times later in the book.

Chapter 6
Conservation of Energy (Bernoulli Balance)

Abstract In the absence of other, non-mechanical forms of energy transferred from/to the fluid system, conservation of mechanical energy is a direct consequence of conservation of momentum and does describe a separate physical principle. Nevertheless, the law of conservation of mechanical energy provides a different perspective on the phenomena occurring during fluid motion. Under some specific conditions and absence of energy dissipation, conservation of energy can be stated in a particularly simple form, known as the Bernoulli balance, which ensures an immediate physical interpretation. A few basic applications are then presented based on this formulation. Local phenomena of energy dissipation are then analyzed with a mean to include them as an extension of the Bernoulli balance.

6.1 Equation for Conservation of Mechanical Energy

In a system where the only form of energy is mechanical energy, there are no other physical mechanisms, or physical laws, other than conservation of mass and of momentum that can be included to describe the behavior of the system. The unique forms of energy concern kinetic energy $\frac{1}{2}\rho v^2$ (per unit volume) and potential energy $p + \gamma z$ (per unit volume). The latter again underlining that pressure plays the same role as gravity, which is commonly implicitly included therein for easier writing. The law of conservation of momentum already described the dynamic relationship between velocity (for kinetic energy) and pressure (for potential energy); therefore, in absence of other forms of energy, the conservation of energy must be in accordance with the only possible energetic transformations: from kinetic to potential energy and vice versa, and kinetic energy can be dissipated for friction.

In this case, the law of conservation of energy can be obtained directly from the conservation of momentum. Consider the ith component of the Cauchy Eq. (5.24) for a generic continuum

$$\rho \frac{\partial v_i}{\partial t} + \rho v_j \frac{\partial v_i}{\partial x_j} = f_i - \frac{\partial \mathbb{T}_{ij}}{\partial x_j}.$$

© The Author(s), under exclusive license to Springer Nature Switzerland AG 2022
G. Pedrizzetti, *Fluid Mechanics for Cardiovascular Engineering*,
https://doi.org/10.1007/978-3-030-85943-5_6

Now make the scalar multiplication with the velocity (in index formalism, multiply the ith component of the equation by the same component of velocity and perform summation for $i = 1,2,3$). Using the rule for the derivative of a product, this equation can be rewritten as

$$\frac{\partial}{\partial t}\left(\tfrac{1}{2}\rho v_i v_i\right) + v_j \frac{\partial}{\partial x_j}\left(\tfrac{1}{2}\rho v_i v_i\right) = v_i f_i - v_i \frac{\partial \mathbb{T}_{ij}}{\partial x_j}.$$

In the final form, the general equation for the balance of kinetic energy is

$$\frac{\partial}{\partial t}\left(\frac{1}{2}\rho v^2\right) + \boldsymbol{v} \cdot \nabla\left(\frac{1}{2}\rho v^2\right) = \boldsymbol{v} \cdot \boldsymbol{f} - \boldsymbol{v} \cdot (\nabla \cdot \mathbb{T}); \qquad (6.1)$$

The two terms on the left-hand side are the (Lagrangian) time derivative of the kinetic energy per unit volume, $\tfrac{1}{2}\rho v^2$: the rate of change of the kinetic energy evaluated on the moving fluid element. The fluid kinetic energy can change for the rate of work (the power) due to the other forces acting on the element of fluid. The first terms on the right-hand side is the power of the volume force. In the case of gravity, when $\boldsymbol{f} = -\gamma \boldsymbol{k}$, this can be rewritten as $-\boldsymbol{v} \cdot \gamma \boldsymbol{k} = -\gamma v_z$ which is the rate of change of the potential energy associated with the height of the fluid particle. The second term is the power associated with the surface forces. We can immediately recognize the contribution of pressure therein by substituting $\mathbb{T}_{ij} = p\delta_{ij}$, which gives $-\boldsymbol{v} \cdot (\nabla \cdot \mathbb{T}) = -\boldsymbol{v} \cdot \nabla p$; this is the rate of work made by the pressure gradient, or the rate of change of potential energy (per unit volume) p. The contribution of the viscous term, when we substitute $\mathbb{T}_{ij} = -2\mu \mathbb{D}_{ij}$, can be split into two parts $-v_i \frac{\partial \mathbb{T}_{ij}}{\partial x_j} = 2\mu \frac{\partial v_i \mathbb{D}_{ij}}{\partial x_j} - 2\mu \mathbb{D}_{ij}\mathbb{D}_{ij}$. The first part is the total power of the viscous forces. The second is the rate of work spent to deform the fluid elements, which represents the rate of dissipation of kinetic energy due to viscous friction

$$\varepsilon(t) = 2\mu \mathbb{D}_{ij}\mathbb{D}_{ij} = \mu\left(\frac{\partial v_i}{\partial x_j}\frac{\partial v_i}{\partial x_j} + \frac{\partial v_i}{\partial x_j}\frac{\partial v_j}{\partial x_i}\right), \qquad (6.2)$$

which is strictly a positive quantity, because given by the sum of squares, thus unavoidably leading to a reduction of the kinetic energy.

6.2 Bernoulli Energy Balance

An alternate expression for the conservation of energy that is of immediate interpretation can be obtained directly from the Navier–Stokes equation under some specific hypotheses.

As a first hypothesis, assume that volume forces are zero or they are conservative forces such that they can be expresses as the gradient of a force potential, like the gravitational force $\boldsymbol{f} = \nabla(-\gamma z)$. In this case, and without loss of generality, volume forces can be included in the pressure term that has the form of a gradient as well. Now make the strong hypothesis of considering that the fluid behaves like an ideal fluid with zero viscosity such that viscous energy dissipation is absent. The equation governing the motion of an ideal fluid is the Euler Eq. (5.36)

$$\frac{\partial \boldsymbol{v}}{\partial t} + \boldsymbol{v} \cdot \nabla \boldsymbol{v} = -\frac{1}{\rho} \nabla p.$$

In order to move forward, the second term on the left-hand side must be rewritten in an alternate form. To show this, consider the x-component of that term in Cartesian coordinates

$$\boldsymbol{v} \cdot \nabla \boldsymbol{v}|_x = v_x \frac{\partial v_x}{\partial x} + v_y \frac{\partial v_x}{\partial y} + v_z \frac{\partial v_x}{\partial z},$$

now add and subtract the same quantity and write

$$\boldsymbol{v} \cdot \nabla \boldsymbol{v}|_x = v_x \frac{\partial v_x}{\partial x} + v_y \frac{\partial v_x}{\partial y} + v_z \frac{\partial v_x}{\partial z} + v_y \frac{\partial v_y}{\partial x} - v_y \frac{\partial v_y}{\partial x} + v_z \frac{\partial v_z}{\partial x} - v_z \frac{\partial v_z}{\partial x}.$$

The first, fourth, and sixth terms can be grouped as derivative of squares, then evidence v_y from the second and fifth and v_z from third and seventh

$$\boldsymbol{v} \cdot \nabla \boldsymbol{v}|_x = \frac{1}{2} \frac{\partial}{\partial x} \left(v_x^2 + v_y^2 + v_z^2 \right) + v_y \left(\frac{\partial v_x}{\partial y} - \frac{\partial v_y}{\partial x} \right) + v_z \left(\frac{\partial v_x}{\partial z} - \frac{\partial v_z}{\partial x} \right).$$

The first term is the derivative of the square of the modulus of the velocity $v^2 = v_i v_i = v_x^2 + v_y^2 + v_z^2$; then notice that the terms in parenthesis are the cross-components of the vorticity vector, and we can eventually write.

$$\boldsymbol{v} \cdot \nabla \boldsymbol{v}|_x = \frac{\partial}{\partial x} \frac{v^2}{2} - v_y \omega_z + v_z \omega_y;$$

which in vector forms reads

$$\boldsymbol{v} \cdot \nabla \boldsymbol{v} = \nabla \frac{v^2}{2} - \boldsymbol{v} \times \boldsymbol{\omega}. \tag{6.3}$$

Insertion of the equivalence (6.3) into the Euler equation permits to rewrite it in the following alternate form:

$$\frac{\partial \boldsymbol{v}}{\partial t} + \nabla \left(\frac{v^2}{2} + \frac{p}{\rho} \right) = \boldsymbol{v} \times \boldsymbol{\omega}. \tag{6.4}$$

For the property of the cross product, the term on the right-hand side is perpendicular to both velocity and vorticity. Let us now project the vector Eq. (6.4) in the direction of a streamline, which is a direction that is at every point parallel to the velocity vector. To this aim take the scalar product of every term in (6.4) with the unit vector $s = v^{-1}v$, to get (after multiplication with density)

$$\rho \frac{\partial v}{\partial t} \cdot s + \frac{\partial}{\partial s}\left(\frac{1}{2}\rho v^2 + p\right) = 0. \tag{6.5}$$

This equation tells that, in an ideal fluid, in the absence of non-conservative forces, the total mechanical energy (per unit volume) given by the sum of kinetic energy $\frac{1}{2}\rho v^2$ and potential energy p (which includes the potential of gravitational and other conservative forces, if present), can vary along a streamline only when the flow accelerates/decelerates along that direction.

This finding can be made more explicit by integration of Eq. (6.5) between two points, point 1 and point 2, along one streamline

$$p_1 + \frac{1}{2}\rho v_1^2 = p_2 + \frac{1}{2}\rho v_1^2 + \rho \int_1^2 \frac{\partial v}{\partial t} \cdot ds, \tag{6.6}$$

which expresses the conservation of the mechanical energy (per unit of volume) between two points along a streamline. Equation (6.6) represents the *Bernoulli theorem* or *Bernoulli balance* and states that when viscous energy dissipations are negligible, the total mechanical energy is conserved along a streamline net of the last term (inertia) that is the energy spent to accelerate the fluid or acquired during its deceleration.

For an immediate interpretation, it is common to define the total head

$$H = \frac{v^2}{2g} + \frac{p}{\gamma} = \frac{v^2}{2g} + h, \tag{6.7}$$

which is the height that expresses the total mechanical energy per unit weight as the sum of kinetic energy $\frac{v^2}{2g}$ plus the potential energy h, which is the static head previously defined by Eq. (2.6) in Sect. 2.1. Using the definition (6.7), the Bernoulli balance (6.6) can be written as

$$H_1 = H_2 + \frac{1}{g} \int_1^2 \frac{\partial v}{\partial t} \cdot ds; \tag{6.8}$$

stating that the total head can vary along a streamline only when there is a variation of fluid inertia along that path.

The special simple case when the inertial term can be neglected takes particular relevance for numerous applications. This simplification applies in general to flows

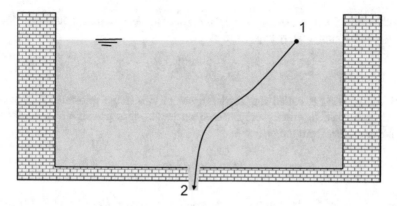

Fig. 6.1 Flow existing from the bottom of a reservoir

in a steady state, having $\frac{\partial v}{\partial t} = 0$; however, it also applies to unsteady flows at the instant of maximum or minimum velocity when $\frac{\partial v}{\partial t} \cong 0$ and to the motion averaged during the period of a periodic flow. In this case, the last term in (6.8) is zero and the total head (6.7) is conserved along a streamline

$$H_1 = H_2. \tag{6.9}$$

This situation presents only the transformation of kinetic energy into potential energy (pressure) and vice versa.

In order to show the utility of the Bernoulli theorem in some simple applications, consider the case of a reservoir with an exit at its bottom as shown in Fig. 6.1, assuming the reservoir large enough that we can consider the flow as approximately stationary.

Take a streamline starting from the free surface and reaching the outflow and state the Bernoulli balance (6.9) using the definition (6.7) of the total head writing the gravitational term explicitly

$$\frac{v_1^2}{2g} + \frac{p_1}{\gamma} + z_1 = \frac{v_2^2}{2g} + \frac{p_2}{\gamma} + z_2. \tag{6.10}$$

Pressure is equal to the atmospheric pressure at point 1 on the free surface and at point 2 that is a unidirectional jet surrounded by atmospheric pressure. As the reservoir is large, we can neglect the velocity (square) on the free surface with respect to that at the outlet. The previous balance simplifies in

$$z_1 = \frac{v_2^2}{2g} + z_2.$$

Indicating with $h = z_1 - z_2$, the total head bearing on the exit, the outflow velocity $v = v_2$ can be immediately expressed as

$$v = \sqrt{2gh}. \tag{6.11}$$

Velocity (6.11) is called the Torricelli velocity; it is the free-fall velocity of a particle subjected to gravity only. Based on (6.11), it is possible to estimate the discharge exiting from the orifice as

$$Q = C_c A \sqrt{2gh},$$

where A is the orifice area and C_c is the coefficient of contraction that accounts for the contraction of the cross-section of the existing jet, which for a sharp edge is about $C_c \cong 0.6$.

In a steady flow, in general, the change of fluid kinetic energy is balanced by a change in pressure. Therefore, for example, the fluid velocity increases in a horizontal converging vessel and the pressure decreases accordingly to the conservation of the total head; vice versa, in an expanding vessel, velocity decreases downstream while pressure increases. Similarly, when you have a steady jet, with velocity v, that impacts on a solid surface, the stagnation point on the solid surface experiences an overpressure

$$\Delta p = \frac{1}{2} \rho v^2;$$

because the entire incoming kinetic energy is transformed into an increase of pressure.

The Bernoulli balance is at the base of an important velocity measurement instrument called *Pitot tube* that is shown in Fig. 6.2. The Pitot tube is a small tube with a bullet-like leading edge facing the incoming stream. It is made of two concentric chambers: the inner chamber communicates to the outside from an opening at the front tip; the outer chamber communicates to the outside through openings on the lateral side. Then, the two chambers end internally with a differential manometer that reports their pressure difference.

With reference to the sketch in Fig. 6.2, we can apply the Bernoulli balance, under steady conditions, separately for the two chambers along two streamlines both starting from two points upstream that are very close to each other, thus have the same velocity v and pressure p (point 0). The first streamline selected is ending precisely to the stagnation point (point 1) in front of the tube, the other is one passing to the side near the lateral holes (point 3).

Consider first the path starting from the upstream point 0, passing through point 1, and ending to point 2 on one side of the differential manometer. Apply the Bernoulli balance between 0 and 1,

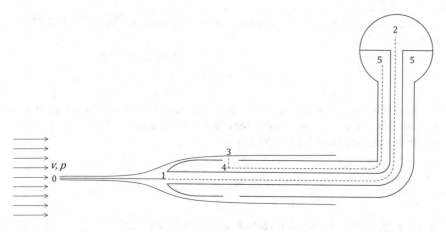

Fig. 6.2 Pitot tube, sketch for calculations

$$\frac{p}{\rho} + \frac{v^2}{2} = \frac{p_1}{\rho} + \frac{v_1^2}{2}.$$

Velocity is zero at the stagnation point 1 and we obtain that pressure measured in 1 is equal to the upstream pressure augmented by the kinetic energy that is transformed into pressure at the stagnation point

$$p_1 = p + \rho \frac{v^2}{2}.$$

Then, inside the tube, the fluid is at rest and the laws of fluid statics hold. Ignoring gravity (without loss of generality, because it can be included in pressure), we have that

$$p_2 = p_1 = p + \rho \frac{v^2}{2}.$$

Now consider the path starting from the upstream point 0, passing through points 3 and 4, and ending to point 5 on the other side of the differential manometer. The Pitot tube is small enough that it does not disturb appreciably the fluid flow and we can assume that next to the tube $v_3 = v$ and $p_3 = p$. The Bernoulli balance cannot be applied between points 3 and 4 because there is no streamline connecting the two. However, the path from point 3 to point 4 moves transversally to the streamlines and we can apply the law of statics (5.38) transversal to the direction of flow. Then, once entered into the tube the same law of statics applies in the fluid at rest. This gives a constancy of pressure from the outside up the manometer

$$p_5 = p_4 = p_3 = p.$$

From these formulas, the pressure values measured at the two sides of the manometer are

$$p_2 = p + \rho \frac{v^2}{2}, \quad p_5 = p;$$

the former is often called the dynamic pressure, because it is the ambient pressure increased by the kinetic head; the latter is the static ambient pressure. The pressure difference reading from the manometer is

$$\Delta p = p_2 - p_5 = \rho \frac{v^2}{2},$$

which is immediate to transform into a velocity measurement

$$v = \sqrt{2 \frac{\Delta p}{\rho}}. \tag{6.12}$$

The Pitot tube has an important applied relevance because it provides a measurement of velocity based on mechanical principles. It works without the need for neither an external source of energy like electricity nor digital post-processing of data. It thus equips most aircrafts and boats providing an independent velocity measurement to rely on under any circumstances up to the case of failure of electric support.

It is worth to underline that the pressure difference in Eq. (6.12) can be rewritten as a difference of static head Δh, such that $\Delta p = \gamma \Delta h$. If the two chambers were connected to fluid-filled vertical ducts, the internal chamber would show a level of the free surface that is Δh higher than the external chamber. Such difference corresponds to the additional pressure imputable to the kinetic energy

$$v = \sqrt{2g \Delta h}. \tag{6.13}$$

As a raw example, consider a boat traveling at a speed v. Immerge a small tube vertically and curve its lower end such that its entrance faces the flow (at sufficient distance to assume the flow undisturbed by the presence of the boat). Such a tube plays the role of the internal chamber of a raw Pitot tube, the fluid therein will rise of a height Δh above the free surface of the surrounding fluid that represents the static head. Measurement of the Δh in the curved tube and Eq. (6.13) allow evaluating the boat velocity through a purely mechanical balance.

The Pitot tube represents the archetype of velocity and pressure measurements used in clinical practice through catheterization. Typically, clinical hemodynamic catheters are used inside the heart chambers or in large vessels and present side openings to measure pressure and possibly front openings to measure velocities.

In cardiovascular pulsatile flows, the simplified, stationary form of the Bernoulli balance can be used during those time instants when the time derivative of velocity is

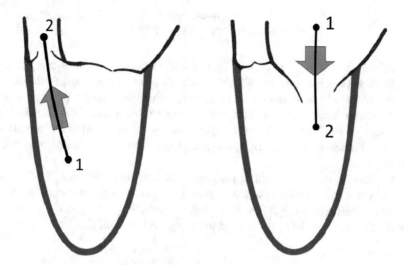

Fig. 6.3 Flow across the aortic valve (left) or mitral valve (right)

zero. In particular, it is commonly employed at the peak of the pulsation to compute the pressure drop across cardiac valves.

With reference to Fig. 6.3 (left), consider the blood flow ejected across the aortic valve during systole. Select a portion of streamline crossing the valve, take the first point inside the ventricle and the second point at the exit of the valve. The same approach can be used for flow across the mitral valve (Fig. 6.3, right), The velocity is at its maximum during the pulsation and the time derivative is approximately zero, thus Bernoulli balance (6.6) reads

$$\frac{p_1}{\rho} + \frac{v_1^2}{2} = \frac{p_2}{\rho} + \frac{v_2^2}{2}.$$

Neglecting the upstream velocity (square) inside the chamber with respect to the velocity at the exit of the valvular tips, the pressure drop $\Delta p = p_1 - p_2$ across the valve can be expressed as a function of the valvular velocity $v = v_2$ as

$$\Delta p = \rho \frac{v^2}{2}. \tag{6.14}$$

The velocity at the exit of the valve can be measured with relative ease, for example, with Doppler ultrasound, and allows having an estimate of the transvalvular pressure drop. Formula (6.14) is dimensionally consistent; for example, when the valve velocity is measured in *m/s* and density in *Kg/m³* then pressure is given in *Pascal*. In clinical practice, it is very common to express this balance in a simplified form, with pressure drop measured in mm_{Hg} and velocity in *m/s*. The transformation from *Pascal* to mm_{Hg} requires a factor 133 Pa/mm$_{Hg}$, and using density $\rho = 1050$ kg/m³ then

$$\Delta p_{[mm_{Hg}]} = \frac{1050}{133 \times 2} v^2_{[\frac{m}{s}]} \cong 4 v^2_{[\frac{m}{s}]}.$$

The simple formula $\Delta p = 4v^2$ often called the *simplified Bernoulli formula* (which we remark is valid only when pressure is measured in mm_{Hg} and velocity in *m/s*) is widely used in clinical cardiology to estimate transvalvular pressure gradients. Given its frequent use, it must be kept in mind that it was obtained under the hypotheses of the Bernoulli balance (ideal flow), with the additional assumptions that the upstream velocity is negligible and it is valid at the instant of maximum velocity.

The pressure drop evaluated at the instant of maximum velocity (6.14) is not necessarily the maximum pressure drop during the period of systolic outflow across the aortic valve (or diastolic inflow through the mitral valve). At a generic instant, the time derivative cannot be neglected and the complete Bernoulli balance (6.6) applies

$$\Delta p = p_1 - p_2 = \frac{1}{2}\rho(v_2^2 - v_1^2) + \rho \int_1^2 \frac{\partial v}{\partial t} \cdot ds,$$

which can be rewritten as.

$$\Delta p = \frac{1}{2}\rho(v_2^2 - v_1^2) + \rho \frac{\partial \overline{v_{12}}}{\partial t} L_{12},$$

where L_{12} is the distance traveled between the two points and $\overline{v_{12}}$ is the velocity averaged along that path. Under the assumption that the upstream velocity is negligible, $v_1 \ll v_2 = v$, and that the average velocity is approximately $\overline{v_{12}} \cong \frac{1}{2}v$, then we can approximate the unsteady pressure drop by

$$\Delta p = \frac{\rho}{2}v^2 + \rho \frac{\partial \overline{v_{12}}}{\partial t} L_{12} \cong \frac{\rho}{2}v^2 + \frac{\rho}{2}\frac{\partial v}{\partial t} L_{12}. \qquad (6.15)$$

The first term is the pressure drop due to the transformation of pressure into kinetic energy, the second is the energy stored into inertia. Typically, the two terms are comparable in magnitude and present different time phases. The former is in phase with velocity and dominates about the instants of maximum velocity; the latter is in quadrature of velocity and dominates during the acceleration/deceleration periods (Firstenberg et al. 2000; Tonti et al. 2001).

6.3   Bernoulli Balance with Dissipation: Localized Energy Losses

Among the hypotheses of the Bernoulli balance, the one that can be unrealistic under several conditions is the assumption of an ideal fluid. Real fluids are not ideal and some form of viscous dissipation is always present.

In general, when the flow in a vessel passes across a reduction of area (like a valve or a vessel narrowing like a stenosis) the flow accelerates across the reduced section and decelerates afterward. Thus, potential energy (pressure) transforms into kinetic energy at the constriction where the pressure reaches smaller values, then the kinetic energy transforms back into pressure at the expansion. However, also a net loss of energy occurs along this short track. Consider, for example, a straight duct with a reduction of diameter during a brief portion of its length. After the constriction, velocity (and kinetic energy) returns to the value it had before the constraint as dictated by mass conservation; however, pressure does not get back to its initial value and displays a net reduction. This reduction is due to the energy lost for friction along the short tract presenting narrowing and expansion. We consider these as *localized* energy losses, because they occur in consequence of a local disturbance to the flow.

In general, dealing with friction requires the use of the Navier–Stokes equation that introduces several complexities in the analysis. However, it is sometimes feasible to extend the formulation of the Bernoulli balance (6.6) by simply adding an extra term testifying to the presence of energy dissipation

$$p_1 + \rho \frac{v_1^2}{2} = p_2 + \rho \frac{v_2^2}{2} + \rho \int_1^2 \frac{\partial \boldsymbol{v}}{\partial t} \cdot d\boldsymbol{s} + \Delta E, \qquad (6.16)$$

including an explicit term ΔE accounting for the amount of energy lost traveling from point 1 to point 2. Equation (6.16) maintains the same form of the Bernoulli balance although it contains an additional term that is in principle unknown. However, under some circumstances, the pressure loss can be expressed in simple form as a percentage of the available kinetic energy. In that case, the generalized balance (6.16) can be used in the same way as the normal Bernoulli balance.

An exemplary case where the energy losses can be evaluated with relative ease is the case of a sudden expansion and rigid walls as sketched in Fig. 6.4.

To this aim, the balance of momentum

$$I_x + M_x = \Pi_x$$

should be written for the cylindrical volume of cross-area A_2 and length L, indicated with a dashed line in Fig. 6.4, starting adjacent to the expansion and ending in a far-enough downstream section where the flow can be assumed to be returned to unidirectional.

Fig. 6.4 Sketch for evaluating energy loss in a sudden expansion

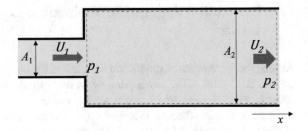

The inertial term is

$$I_x = \int_V \frac{\partial \rho v_x}{\partial t} dV = \rho \int_L \int_{A_2} \frac{\partial v_x}{\partial t} dA dx = \rho \frac{dU_2}{dt} A_2 L.$$

The flux of momentum occurs across the open part, of size A_1, of the upstream section and across the entire downstream section of area A_2

$$M_x = -\rho U_1^2 A_1 + \rho U_2^2 A_2;$$

assuming that the velocity is approximately uniform over both cross-sections ($\beta \cong 1$).

The pressure term pushes upwards on the downstream surface of area A_2 where pressure is p_2; it also pushes, downward, on the entire upstream surface of area again equal to A_2, here pressure is equal to p_1 on the open part and it remains approximately constant on the closed part where flow is about stagnating, thus the law of static applies

$$\Pi_x = p_1 A_2 - p_2 A_2.$$

Summing up these three terms and dividing by A_2 we obtain

$$p_1 - p_2 = \rho \frac{dU_2}{dt} L - \rho U_1^2 \frac{A_1}{A_2} + \rho U_2^2. \tag{6.17}$$

The balance (6.16) can be rewritten making the dissipation term explicit

$$\Delta E = p_1 - p_2 + \rho \frac{v_1^2}{2} - \rho \frac{v_2^2}{2} - \rho \int_1^2 \frac{\partial \mathbf{v}}{\partial t} \cdot d\mathbf{s}; \tag{6.18}$$

then, assuming the flow sufficiently uniform we can exchange velocity and cross-section average velocity and rewrite (6.18)

$$\Delta E = p_1 - p_2 + \rho \frac{U_1^2}{2} - \rho \frac{U_2^2}{2} - \rho \frac{\partial U_2}{\partial t} L. \tag{6.19}$$

Now substitute the pressure difference (6.17) into (6.19) to get

$$\Delta E = \rho \frac{U_2^2}{2} + \rho \frac{U_1^2}{2} \left(1 - 2 \frac{A_1}{A_2} \right);$$

that can be rewritten in terms of one velocity only

$$\Delta E = \left(1 - \frac{A_1}{A_2} \right)^2 \rho \frac{U_1^2}{2} = \eta \rho \frac{U_1^2}{2}, \ \eta = \left(1 - \frac{A_1}{A_2} \right)^2. \tag{6.20}$$

Equation (6.20) describes the loss of energy (per unit volume) in a sharp enlargement. It tells that energy losses are given by a fraction η of the incoming kinetic energy, while the remainder is transformed into potential energy (i.e., pressure). The entity of such a fraction depends on the degree of the expansion. In the limit case of very large expansion, $A_2 \gg A_1$, then $\eta \cong 1$, the incoming kinetic energy is unable to significantly affect the very wide downstream reservoir and the entire incoming kinetic energy is lost.

The result (6.20), despite some simplificative assumptions, is very instructive because it teaches that localized energy losses can be in general expresses as a fraction of the available kinetic energy.

$$\Delta E = \eta \rho \frac{U^2}{2}, \tag{6.21}$$

where the dimensionless dissipation coefficient η depends on the degree of disturbance created on the streaming flow. In most cases, the dissipation coefficient cannot be easily expressed by explicit formulas like (6.20). However, its numerical value was determined experimentally in numerous situations of practical interest and can be often found in the literature.

Part III
Fundamentals for Mostly Unidirectional Flow

Chapter 7
Unidirectional Flow in Rectilinear Vessels

Abstract The equations governing fluid motion allow a mathematical solution in a few simple conditions only. The unidirectional flow in rectilinear geometry is one important group of such classes, where the exact solution for a cylindrical duct represents a reference for blood flow in straight vessels far enough from the entrance and in absence of other geometrical disturbances. The analysis is preceded by the introduction of the concept of the boundary layer, which describes the behavior of fluid motion in the vicinity of a wall where viscous adherence represents the dominant phenomenon. Then we derive the classical solution of the Navier–Stokes equations for steady flow in a cylindrical duct (Poiseuille flow); the analysis is then extended to pulsatile flows where a few important dimensionless numbers naturally appear for characterizing the possible behaviors of fluid motion.

7.1 Boundary Layer

Viscosity is the unique property associated with the development of friction and energy dissipation in fluids governed by the Navier–Stokes equation; particularly, in the vicinity of solid walls in association to the phenomenon of viscous adherence. In order to understand the role of viscosity in more depth, let us analyze the details of the flow near the boundary in the simple case of a perfectly flat, rigid wall aligned with the flow. This configuration is specifically designed to explore the role of viscosity because in absence of adherence the flow would slip over the wall without resistance and would be unaffected by the presence of the wall.

Consider the flow directed along the x-direction of a Cartesian set of coordinates, with the wall set at $y = 0$, and neglect the velocity and its variation along the transversal z-component (two-dimensional flow). The stream-wise component of the Navier–Stokes equation for this case is

$$\frac{\partial v_x}{\partial t} + v_x \frac{\partial v_x}{\partial x} + v_y \frac{\partial v_x}{\partial y} = -\frac{1}{\rho}\frac{\partial p}{\partial x} + \nu\left(\frac{\partial^2 v_x}{\partial x^2} + \frac{\partial^2 v_x}{\partial y^2}\right). \tag{7.1}$$

We have seen that the kinematic viscosity in front of the last term is a small number. Therefore, the viscous term is often negligible locally. However, viscosity has a

G. Pedrizzetti, *Fluid Mechanics for Cardiovascular Engineering*,
https://doi.org/10.1007/978-3-030-85943-5_7

fundamental role in the proximity of solid boundaries because it is associated with the boundary condition of adherence, which applies irrespective of the value of viscosity. As a result, viscosity unavoidably affects the flow near a solid boundary (because of adherence) while its influence is expected to become progressively smaller moving away from it. In other terms, there is always a region next to the wall boundary, which is called *boundary layer*, where the role of viscosity is unavoidably important. As a possible definition, in mathematical terms, the boundary layer is the region next to the boundary where the entity of the viscous term in the Navier–Stokes equation is comparable with other terms.

Consider a uniform unidirectional flow, with a velocity equal to U at every point, that encounters a plane surface of negligible thickness. As shown in Fig. 7.1, when the incoming uniform profile gets in contact with the surface the velocity at the surface goes to zero because of adherence. As the fluid travels downstream, the slower fluid elements close to the boundary decelerate those immediately above thus extending the influence of adherence for a thickness of fluid over the surface. This process continues and the thickness influenced by the viscous adherence increases downstream. Roughly speaking, the flow field can be divided into an external flow, not reached by the influence of adherence, and a boundary layer that is directly affected by viscosity.

The thickness of the boundary layer is indicated with $\delta(x)$ and it increases downstream. The upper limit of the boundary layer is not identified by a definite edge because the influence of viscosity decreases progressively with height. Nevertheless, the order of magnitude of $\delta(x)$ can be obtained by estimating the order of magnitude of the different terms in the Navier–Stokes Eq. (7.1) and applying the definition that the thickness bounds the region where the viscous term is comparable with the others.

Let us discuss the various terms in Eq. (7.1). First, the time derivative can be ignored because the flow is steady. For estimating the magnitude of the second term, the longitudinal transport term, at a point x, we can take one velocity upstream, say at $x = 0$, which is equal to U and one downstream, say at a distance $2x$ inside the boundary layer, which is a fraction of U, say κU. Here κ is a number smaller than 1, but still a finite fraction of 1 and not infinitesimal (in order of magnitude arguments,

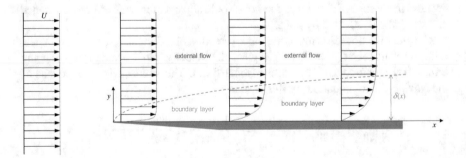

Fig. 7.1 Boundary layer development on a flat plate

κ is said to of the order of magnitude of 1). Thus, when the value of x is small, the velocity derivative at x can be roughly estimated by the difference between the velocity at $2x$ and at 0 divided by the distance, while the velocity therein is estimated as the mean value between the same quantities

$$v_x \frac{\partial v_x}{\partial x} \sim \frac{\kappa U + U}{2} \cdot \frac{\kappa U - U}{2x} = \frac{\kappa^2 - 1}{2} \frac{U^2}{x} \sim \frac{U^2}{x}; \tag{7.2}$$

where the symbol $\sim$ stands for "of the order of magnitude of" and coefficients, that are about the order of unity, are omitted to keep the attention on the dependencies in terms of order of magnitude.

When estimating the third term (the second transport term), the order of magnitude of the transversal velocity, v_y, can be devised by the continuity equation that in 2D reads

$$\frac{\partial v_x}{\partial x} + \frac{\partial v_y}{\partial y} = 0. \tag{7.3}$$

The x-derivative can be estimated as above; the y-derivative can be roughly given by the unknown value v_y, minus the zero value at the wall, divided by the boundary layer thickness. Thus (7.3) tells

$$\frac{U}{x} \sim \frac{v_y}{\delta}. \tag{7.4}$$

The velocity v_y for the second transport term in (7.1) is taken by (7.4) and the y-derivative is estimated by the velocity U outside the boundary layer, divided by the thickness δ

$$v_y \frac{\partial v_x}{\partial y} \sim v_y \frac{U}{\delta} \sim \frac{U \delta}{x} \frac{U}{\delta} \sim \frac{U^2}{x} \tag{7.5}$$

that turns out to be of the same order of the other transport term (7.2). Let us ignore for the moment the pressure term in (7.1), as the boundary layer develops even in absence of pressure gradient thus it should not play a key role.

Following the same lines, the viscous term is

$$\nu \left(\frac{\partial^2 v_x}{\partial x^2} + \frac{\partial^2 v_x}{\partial y^2} \right) \sim \nu \left(\frac{U}{x^2} + \frac{U}{\delta^2} \right) \sim \nu \frac{U}{\delta^2}, \tag{7.6}$$

where we ignored the first term in parenthesis with respect to the second because we expect δ to be small, to be more precise it is assumed that $\delta \ll x$.

In the boundary layer, the viscous term is of the same order of magnitude as the other terms. Equating (7.6) with (7.2) or (7.5) we obtain

$$\nu\frac{U}{\delta^2} \sim \frac{U^2}{x};$$

and therefore the order of magnitude of the boundary layer thickness turns out to be

$$\delta \sim \sqrt{\nu\frac{x}{U}}.\tag{7.7}$$

The estimate (7.7) shows that the thickness of the boundary layer grows with the square root of the downstream distance. The boundary layer is thin when viscosity is small and gets thinner when velocity is higher. The exact coefficient that is in front of the square root of (7.7) depends on the specific definition of boundary layer thickness. Some texts suggest setting the edge of the boundary layer where velocity differs of a predefined small percentage to the external velocity, others use the velocity square (momentum). In any case, Eq. (7.7) has a general validity and the coefficient in front varies from about 2–5, depending on the exact definition of δ and on the specific situation under analysis.

Let us try to understand further the origin of the boundary layer thickness. The mathematical structure of the viscous term in the Navier–Stokes equation corresponds to a phenomenon of diffusion. It represents the diffusion of a disturbance to velocity (set to zero by adherence) from the wall into the bulk flow. Indeed, pure diffusion of a whatsoever field $f(t, y)$ along the direction, say y, is described by the diffusion equation

$$\frac{\partial f}{\partial t} = D\frac{\partial^2 f}{\partial y^2},\tag{7.8}$$

where D is the diffusion coefficient (in Navier–Stokes corresponding to the kinematic viscosity).

The connection between diffusion and boundary layer development is immediate considering the dual problem of what is discussed above, that of a fluid initially at rest over an infinite plate that is abruptly set in motion with velocity U. In this case, the origin of boundary layer development occurs at an initial time, say $t = 0$, and it is uniform in space; whereas in the previous example boundary layer developed from a spatial position, $x = 0$, and did not change in time.

When the boundary layer is uniform along x and grows in time, the Navier–Stokes equation reads

$$\frac{\partial v_x}{\partial t} = \nu\frac{\partial^2 v_x}{\partial y^2},\tag{7.9}$$

which is exactly a diffusion equation like (7.8). This is a linear partial differential equation of parabolic type that was largely investigated in the past. The solution to (7.9) with boundary condition $v_x(0, t) = U$ is the error function

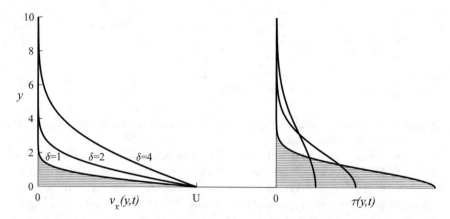

Fig. 7.2 Velocity (left) and shear stress (right) above a moving wall

$$v_x(y, t) = U - \frac{U}{\sigma(t)} \int_0^y e^{-\frac{1}{2}\left(\frac{s}{\sigma(t)}\right)^2} ds \tag{7.10}$$

with

$$\sigma = \sqrt{2vt} \tag{7.11}$$

representing the width of the velocity profile which starts from U at the wall and decreases away from the wall reaching $v_x \approx 0.005U$ at $y = 2\sigma$. The velocity profile (7.10) is shown in Fig. 7.2, where we defined the thickness of the boundary layer by

$$\delta(t) = 2\sigma = 2\sqrt{2vt}. \tag{7.12}$$

The same solution (7.10)–(7.11) can be reached by the perspective of shear rate $\frac{\partial v_x}{\partial y}$. Consider that the wall motion creates a difference in velocity from the value at the wall equal to U to zero infinitely above. This velocity difference is equal to the integral of the velocity gradient

$$U = \int_0^\infty \frac{\partial v_x}{\partial y} dy, \tag{7.13}$$

which is produced by a wall shear stress due to adherence that is $\tau_0 = \mu \left.\frac{\partial v_x}{\partial y}\right|_{y=0}$. When the wall sets in motion, at $t = 0$, the shear stress at the wall is ideally infinite although its distribution $\tau(y)$ has a finite integral (7.13). Then the wall shear stress progressively decreases in time to ensure the same velocity difference while it diffuses away from the wall. The propagation of shear stress is again a diffusion process ruled by the diffusion Eq. (7.8) for $f = \frac{\partial v_x}{\partial y}$. The solution to (7.8) with the constraint (7.13) is the well-known Gauss function

$$\frac{\partial v_x}{\partial y} = \frac{U}{\sigma(t)} e^{-\frac{1}{2}\left(\frac{y}{\sigma(t)}\right)^2};$$

(7.14)

in association with the width $\sigma(t)$ given by (7.11). Then the solution (7.10) can be recovered by integrating (7.14).

We have shown that the development of the viscous boundary layer is simply a phenomenon of diffusion of shear from the wall with thickness given by (7.12). To complete the parallel between the two cases, it is instructive to apply this physical interpretation to the original situation of a steady flow over a plane wall leading to the expression (7.7). In that case, the spatial variation under steady conditions can be transformed into the same diffusion problem by considering an observed moving with velocity U. This observer starts at $t = 0$ from the edge of the plate where $\delta = 0$ and reaches at time t the position $x = Ut$ where the thickness is given by (7.7). Thus, the moving observer sees a boundary layer thickness that grows with time $t = \frac{x}{U}$ with the same law given by (7.12) and the coefficient in (7.7) is defined by

$$\delta = 2\sqrt{2v\frac{x}{U}}.$$

(7.15)

The parallel between boundary layer development and viscous diffusion process is fundamental for the physical interpretation of adherence and of its influence in different conditions.

In a closed conduit, the boundary layer cannot grow indefinitely, because its diffusive growth eventually saturates the available space. Therefore, in a vessel of diameter D, the boundary layer terminates its growth when $\delta \approx D/2$ at a position $x = x_E$. Inserting these values in formula (7.15) it is possible to estimate the length of an entry region involved in boundary layer development in a duct by

$$x_E \approx \frac{1}{32}\frac{UD^2}{v} = \frac{Re}{32}D,$$

(7.16)

where $Re = \frac{UD}{v}$ is the Reynolds number. The boundary layer grows as by (7.15) from the start of the duct, at $x = 0$, to reach an approximately steady thickness starting from about x_E; afterwards, for $x > x_E$, the flow can be assumed as fully developed (spatially uniform) and it does not vary appreciably with the distance x.

These estimates are obtained under the assumption of steady flow and the unsteady case will be considered later. For providing estimates in real arteries, let us assume that this approach is reliable when applied to the time-averaged flow. In the aorta, the mean velocity is about 50 cm/s and the diameter is about 3 cm; the entry flow length is over 100 diameters (a few meters), therefore the flow is never fully developed.

Vice versa, in small arteries the boundary layer fills the entire vessel after less than one diameter downstream the entrance.

	U	D	x_E/D	
Aorta	50 cm/s	3 cm	142	Never fully developed
Mid-vessel	10 cm/s	1 cm	10	
Small-vessel	5 cm/s	2.5 mm	1	Immediately fully developed

7.2 Steady Uniform Planar Flows

Navier–Stokes equation cannot be solved in general; however, a solution can be found under special simple conditions that may present applied relevance. We present here the analytical solution of the Navier–Stokes equation for a few simple cases corresponding to steady (independence on time), unidirectional (velocity vector has one non-zero component) and uniform (independence on the position along the flow). These initial examples will naturally lead to the important case of fluid transported in a cylindrical vessel.

(i) *Flow induced by a moving surface above a fixed wall (Couette flow)*

With reference to Fig. 7.3, consider two plane surfaces, placed at a distance d, with the upper surface moving with constant velocity U relative to the fixed lower surface.

Make the hypothesis that the flow is unidirectional, $v_y = v_z = 0$; that flow is two dimensional, thus derivatives along z are neglected; that flow is in a steady state; thus, time derivatives are neglected, and that flow is due to the wall motion only without pressure gradient $\frac{\partial p}{\partial x} = 0$. The assumption of unidirectional flow implies, by the continuity equation, that the flow is also uniform

$$\frac{\partial v_x}{\partial x} + \frac{\partial v_y}{\partial y} + \frac{\partial v_z}{\partial z} = 0 \Rightarrow \frac{\partial v_x}{\partial x} = 0.$$

The Navier–Stokes equation in the direction perpendicular to the direction of motion simply states that pressure does not vary along y and z; thus, pressure is

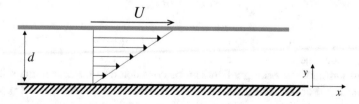

Fig. 7.3 Flow induced by a moving wall

constant everywhere. The only non-zero unknown is the longitudinal component of velocity that is a function of the transversal position, $v_x(y)$. With these assumptions, the Navier–Stokes equation in the direction of the flow (taken as the x-direction) takes the extremely simple form

$$\frac{\partial^2 v_x}{\partial y^2} = 0, \tag{7.17}$$

which must be solved with boundary conditions due to adherence $v_x(0) = 0$ and $v_x(d) = U$.

The solution is immediate to find

$$v_x(y) = U\frac{y}{d}; \tag{7.18}$$

the velocity increases linearly from zero at the lower wall to the value of the moving wall as shown in Fig. 7.3. The shear stress is constant

$$\tau = \mu\frac{dv_x}{dy} = \mu\frac{U}{d}; \tag{7.19}$$

as briefly shown in Sect. 1.2.

(ii)   *Flow between parallel walls.*

Consider the flow induced by a pressure gradient between two plane walls, placed at a distance d.

Make the hypothesis that the flow is unidirectional, two-dimensional, and stationary. Following the same argument used in the previous example, the continuity equation implies that the flow is also uniform. When the flow is uniform and unidirectional, the Navier–Stokes equation in the transversal tells that pressure does not vary transversal to the direction of motion and that the transport term is identically zero in the direction of motion. Thus, Navier–Stokes becomes in the direction of motion (taken as the x-direction) is

$$\frac{1}{\rho}\frac{\partial p}{\partial x} = \nu\frac{\partial^2 v_x}{\partial y^2}; \tag{7.20}$$

it must be solved with boundary conditions due to adherence $v_x\left(\pm\frac{d}{2}\right) = 0$, where we have placed the x-axis as located in the mid-line between the two walls.

In this case, the pressure gradient can be considered as the known quantity that forces the flow. For simplicity, call it $\kappa = -\frac{1}{\rho}\frac{\partial p}{\partial x}$, with the minus sign because pressure is higher upstream than downstream to induce a positive velocity. The solution to (7.20), being κ a constant, is immediate to find and it gives the parabolic profile

$$v_x(y) = \frac{\kappa}{2\nu}\left(\frac{d^2}{4} - y^2\right).$$ (7.21)

It is zero at both walls and has a maximum value at the mid-line between the walls. The shear stress corresponding to the parabolic profile (7.21) is linear

$$\tau = \mu\frac{dv_x}{dy} = -\rho\kappa y = \frac{\partial p}{\partial x}y;$$ (7.22)

taking its maximum value $\tau_0 = \pm\rho\kappa\frac{d}{2}$ with an opposite sign on the opposite walls.

7.3 Steady Uniform Flow in a Circular Vessel (Poiseuille Flow)

The previous flow fields were presented just to introduce the case of higher applied relevance of steady uniform flow in a rectilinear vessel with a circular cross-section. This case represents the effective flow that establishes under steady conditions in many actual vessels along the circulation. It thus applies to veins, where flow is approximately steady, as well as to the time-average flow in some arteries or in slowly varying unsteady flows (as explained later in this section).

With reference to Fig. 7.4, make the hypothesis that the flow is unidirectional, axially symmetric (circular symmetry), and stationary. Thus, as we have seen above, the continuity states that the velocity field is also uniform along the direction of the vessel, taken as the x direction. Pressure is constant transversally to the direction of motion and the transport term is identically zero in the direction of motion. The unknown is the stream-wise velocity that varies on the cross-section $v_x(y, z)$.

Under these hypotheses, the Navier–Stokes equation simplifies to

$$\frac{1}{\rho}\frac{\partial p}{\partial x} = \nu\left(\frac{\partial^2 v_x}{\partial y^2} + \frac{\partial^2 v_x}{\partial z^2}\right).$$ (7.23)

The additional assumption of axial symmetry means that the velocity does not vary along the circumference at a given radial distance, and we may write $v_x(y, z) = v_x(r)$

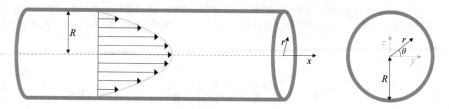

Fig. 7.4 Coordinates and flow in a circular vessel

where $r = \sqrt{y^2 + z^2}$. The viscous term in (7.23) can thus be further simplified passing from Cartesian coordinates (x, y, z) to cylindrical coordinates (x, r, θ) and ignoring the dependence from the angular position θ because of the hypothesis of axial symmetry. For this simplification, the derivatives in Cartesian coordinates y and z are transformed into derivatives with respect to the radial coordinate r by

$$\frac{\partial}{\partial y} = \frac{\partial r}{\partial y} \frac{\partial}{\partial r},$$

$$\frac{\partial^2}{\partial y^2} = \frac{\partial}{\partial y}\left(\frac{\partial r}{\partial y}\frac{\partial}{\partial r}\right) = \frac{\partial^2 r}{\partial y^2}\frac{\partial}{\partial r} + \left(\frac{\partial r}{\partial y}\right)^2 \frac{\partial^2}{\partial r^2}$$

where

$$\frac{\partial r}{\partial y} = \frac{y}{r}, \quad \frac{\partial^2 r}{\partial y^2} = \frac{1}{r} - \frac{y^2}{r^3}.$$

The same can be written by analogy for the z-coordinate to give

$$\frac{\partial^2}{\partial y^2} + \frac{\partial^2}{\partial z^2} = \left(\frac{\partial^2 r}{\partial y^2} + \frac{\partial^2 r}{\partial z^2}\right)\frac{\partial}{\partial r} + \left[\left(\frac{\partial r}{\partial y}\right)^2 + \left(\frac{\partial r}{\partial z}\right)^2\right]\frac{\partial^2}{\partial r^2}$$

$$= \frac{1}{r}\frac{\partial}{\partial r} + \frac{\partial^2}{\partial r^2} = \frac{1}{r}\frac{\partial}{\partial r}\left(r\frac{\partial}{\partial r}\right).$$

In cylindrical, axially symmetric coordinates, Eq. (7.23) can thus be rewritten as

$$\frac{1}{\rho}\frac{\partial p}{\partial x} = \nu\frac{1}{r}\frac{\partial}{\partial r}\left(r\frac{\partial v_x}{\partial r}\right); \tag{7.24}$$

Equation (7.24) can be solved for the unknown velocity profile $v_x(r)$, in correspondence to a given pressure gradient that represents the driving force to the flow, $\kappa = -\frac{1}{\rho}\frac{\partial p}{\partial x}$.

The adherence boundary condition in this case is $v_x(R) = 0$, and we notice that, differently from the case between two walls discussed above, there is only one boundary condition for the second-order differential Eq. (7.24). This is a common consequence of the transformation from Cartesian to cylindrical coordinates because the other boundary at $r = 0$ is not a physical boundary, it is rather a singular point for the presence of the factor $\frac{1}{r}$ arising in the coordinate transformation. Here a regularity condition $|v_x(0)| < \infty$ must be applied and it takes the place of the second boundary condition.

Rewrite (7.24) as

$$-\frac{\kappa}{\nu}r = \frac{\partial}{\partial r}\left(r\frac{\partial v_x}{\partial r}\right),$$

and integrate over r

$$-\frac{\kappa}{2\nu}r^2 = r\frac{\partial v_x}{\partial r} + A,$$

where A is an integration constant, thus

$$-\frac{\kappa}{2\nu}r = \frac{\partial v_x}{\partial r} + \frac{A}{r}.$$

Integrate again over r and get

$$-\frac{\kappa}{4\nu}r^2 = v_x(r) + A\log(r) + B;$$

where B is another integration constant. For the regularity condition $A = 0$, and using the boundary condition at the wall, $B = -\frac{\kappa}{4\nu}R^2$.

The solution eventually is

$$v_x(r) = \frac{\kappa}{4\nu}\left(R^2 - r^2\right); \tag{7.25}$$

which corresponds to a paraboloid solid profile with the maximum velocity at the center of the vessel decreasing to zero at the wall as shown in Fig. 7.5.

The corresponding wall shear stress is

$$\tau_0 = \tau(R) = \mu\left.\frac{dv_x}{dr}\right|_R = -\rho\kappa\frac{R}{2} = \frac{R}{2}\frac{\partial p}{\partial x}, \tag{7.26}$$

which represents the friction exerted by the wall on the flowing fluid.

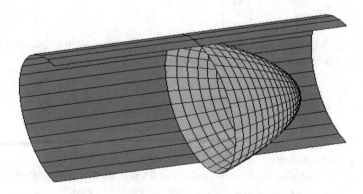

Fig. 7.5 Velocity profile in a circular vessel

The flowing discharge can be computed by integration of (7.25)

$$Q = 2\pi \int_0^R v_x r\, dr = \frac{\pi}{8} \frac{\kappa R^4}{\nu};$$

and the average velocity

$$U = \frac{Q}{\pi R^2} = \frac{1}{8} \frac{\kappa R^2}{\nu}. \tag{7.27}$$

Equation (7.27) is also important as it provides a relationship between the forcing pressure gradient (the cause) and the resulting mean velocity (the effect). Using (7.27) the solution profile (7.25) can be expressed in terms of the mean velocity instead of the pressure gradient

$$v_x(r) = 2U \left(1 - \frac{r^2}{R^2} \right), \tag{7.28}$$

which also shows that the maximum velocity at the center of the duct, $r = 0$, is equal to $2U$, twice the mean velocity and the wall shear stress (7.26) can also be expressed as

$$\tau_0 = -4\mu \frac{U}{R}. \tag{7.29}$$

In this simple situation, where the Navier–Stokes equation could be solved exactly, it is immediate computing the momentum velocity-correction factor β that appeared in (5.7)

$$\beta = \frac{2\pi \int_0^R v_x^2 r\, dr}{\pi R^2 U^2} = 8 \int_0^1 \left(1 - s^2 \right)^2 s\, ds = \frac{4}{3}. \tag{7.30}$$

It is worth to remark that given a flow rate Q, the pressure loss per unit length increases with the fourth power of the vessel diameter

$$\kappa = -\frac{1}{\rho} \frac{\partial p}{\partial x} = \frac{8\nu Q}{\pi R^4};$$

it is therefore natural to recognize that the decrease of the vessel size is accompanied in the vascular network by division of the vessel into multiple smaller vessels each bringing a much smaller discharge to avoid the development of an unphysiologically increase of the wall shear stress in the smaller downstream vessels.

The steady flow solution (7.25) or (7.28) is the result of a balance (7.24) between the force that moves the fluid due to the pressure gradient pushing over the vessel area, $\pi R^2 \frac{\partial p}{\partial x}$, and the viscous friction at the vessel wall that resists to the motion $2\pi R \tau_0$. In other terms, this flow is associated with a continuous pressure loss due to viscous friction.

We have seen in the previous chapter that local energy dissipation represents a fraction of the available kinetic energy. Here too, the *distributed* energy dissipation can be expressed proportional to the available kinetic energy: energy dissipation per unit length is the pressure gradient (kinetic energy is constant) thus

$$-\frac{dp}{dx} = f(Re)\frac{\rho U^2}{2D},\qquad(7.31)$$

where f is a dimensionless friction coefficient (the minus sign is introduced to have positive quantities and a positive friction factor). This formula was previously introduced by dimensional arguments in Eq. (1.14), where it was also shown that the friction coefficient must depend on the Reynolds number, Re. In this case, where we have solved the dynamical equations, it is possible to determine the exact expression of the function $f(Re)$. Recasting the relationship (7.27) obtained from the flow solution, in the form (7.31), the friction factor for Poiseuille flow is

$$f(Re) = \frac{64}{Re},\qquad(7.32)$$

where the definition of the Reynolds number for this situation is

$$Re = \frac{UD}{\nu}.$$

Equation (7.32) shows that the smaller the Reynolds number and the higher are the energy losses due to viscous friction; vice versa, when the Reynolds number is high, viscous losses decrease progressively (asymptotically to zero when $Re \to \infty$). The Reynolds number represents a dimensionless ratio grading the relevance of viscous effects on the flow; therefore, it is of fundamental importance for classifying the type of flow. It represents a ratio between the kinetic energy available to the flow and its ability to dissipate energy. The smaller the Reynolds number the more the flow is a viscous smooth one, its energy is low with respect to the ability to dissipate. The higher the Reynolds number and the more vigorous and energetic the flow. We will see shortly that when the Reynolds number is higher than a certain threshold the flow is so vigorous with respect to its ability to dissipate that a simple viscous mechanism is insufficient to ensure a stable balance. In that case, the Poiseuille flow is unstable and turbulence appears to increase viscous dissipation.

7.4 Oscillatory and Pulsatile Uniform Flow in a Circular Vessel

Blood motion in cardiovascular vessels can be close to steady only in small capillaries and veins. In large arteries, flow typically presents a pulsatile behavior, which can be seen as given by a mean motion plus a fluctuation of a comparable entity. Let us now move forward and consider the solutions of unsteady flows, starting from simple cases and progressing toward more realistic ones.

(i) *Sinusoidal Oscillatory flow*

Consider the case of flow given by an oscillatory pressure gradient of the sinusoidal type

$$\frac{1}{\rho}\frac{\partial p}{\partial x} = \kappa \sin\omega t, \tag{7.33}$$

where the frequency $\omega = \frac{2\pi}{T}$ and T is the period of the oscillation. Under the identical hypothesis used for the Poiseuille flow and only removing the assumption of steady flow, the unknown function is the unsteady velocity $v_x(t, r)$ that obeys the Navier–Stokes equation that in this case reduces to

$$\frac{\partial v_x}{\partial t} + \kappa \sin\omega t = v\frac{1}{r}\frac{\partial}{\partial r}\left(r\frac{\partial v_x}{\partial r}\right); \tag{7.34}$$

with no-slip boundary condition at the wall, $v_x(t, R) = 0$, and regularity condition at $r = 0$. The linearity of Eq. (7.34) tells that that the solution must be time periodic with the same frequency ω of the forcing (7.33) although possibly a different phase (a combination of sine and cosine, or—more properly—an imaginary exponential). The solution of Eq. (7.34) with its boundary conditions can be obtained analytically as (Schlichting 1979)

$$v_x(t, r) = \frac{1}{2}\frac{\kappa}{\omega}\left[1 - \frac{J_0\left(r\sqrt{\frac{-i\omega}{v}}\right)}{J_0\left(R\sqrt{\frac{-i\omega}{v}}\right)}\right]e^{i\omega t} + c.c. \tag{7.35}$$

Formula (7.35) contains the function $J_0(x)$ that is the Bessel function of the first type of order 0; the complex conjugate, *c.c.*, is part of the solution included to ensure that the final quantity is real. For example, the real sinusoidal function is written in terms of complex exponential $\sin\omega t = -\frac{i}{2}e^{i\omega t} + \frac{i}{2}e^{-i\omega t}$ and can be expressed as $\sin\omega t = -\frac{i}{2}e^{i\omega t} + c.c.$ The denominator in solution (7.35) is just a constant to ensure satisfying the boundary condition at $r = R$. The amplitude of the oscillation is given by the ratio $\frac{\kappa}{\omega}$, it is higher for high pressure gradient and for slow oscillations. The entire term in square bracket is a complex number that modifies the phase of the flow along the radial coordinate r.

To better understand this result, solution (7.35) can be preferably expressed in terms of dimensionless parameters as

$$v_x(t, r) = \frac{\kappa}{2\omega}\left[1 - \frac{J_0\left(\frac{r}{R}W\sqrt{\frac{-i\pi}{2}}\right)}{J_0\left(W\sqrt{\frac{-i\pi}{2}}\right)}\right]e^{i2\pi\frac{t}{T}} + c.c.; \qquad (7.36)$$

where the dimensionless coefficient

$$W = R\sqrt{\frac{\omega}{\nu}\frac{2}{\pi}} = \frac{D}{\sqrt{\nu T}} \qquad (7.37)$$

is the *Womersley number* that gives a measure of the degree of unsteadiness of the oscillation.

The Womersley number can be interpreted as the ratio between the vessel diameter and a measure of the thickness of the boundary layer that develops during the period T of the oscillation. Indeed we have seen before that the thickness of a boundary layer reaches in a time t a value proportional $\sqrt{\nu t}$; in this case, the boundary layer is allowed to grow for a time proportional to T; therefore, the denominator of (7.37), $\sqrt{\nu T}$, is a measure of the maximum thickness that the boundary layer can reach during the oscillation. Velocity profiles at different values of the Womersley number are shown in Fig. 7.6. When W is small, the oscillation is slow, the boundary layer has the time to fill the entire vessel and the flow is a sequence of velocity profiles close to the Poiseuille type that is in phase with the pressure gradient because of the linear relationship between velocity and pressure gradient in Poiseuille solution. On the opposite end, when W is large, the oscillation is rapid, the viscous adherence does not have enough time to affect the internal regions of the wall. The viscous boundary layer is limited to a thin region near the wall while the center of the vessel moves nearly as a flat profile with the marginal influence of viscosity. The viscous layer

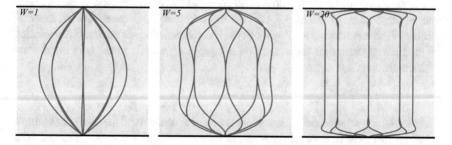

Fig. 7.6 Oscillatory velocity profile in a circular vessel at different Womersley number

near the wall is in phase with the external forcing and gets progressively out of phase away from the wall because the pressure gradient here balances with velocity–time derivative rather than velocity itself.

In mathematical terms, for large W, or sufficiently away from the wall out of the viscous layer, the velocity is not influenced by viscosity and the solution can be obtained neglecting the viscous term in Eq. (7.34), resulting in the asymptotic solution

$$v_x(t,r) = \frac{\kappa}{\omega}\cos\omega t = \frac{\kappa}{2\omega}e^{i\omega t} + c.c., \qquad (7.38)$$

which corresponds to the limit of solution (7.36) when the ratio between Bessel function tends to zero. In the opposite condition, when W is very small, or very close to the wall, the inertial term in (7.34) is negligible with respect to the viscous term and the asymptotic solution is given by (7.25) that can be recast using the forcing (7.33) to give

$$v_x(t,r) = -\frac{\kappa}{\omega}\frac{\pi}{8}W^2\left(1-\frac{r^2}{R^2}\right)\sin\omega t = \frac{\kappa}{2\omega}\frac{i\pi}{8}W^2\left(1-\frac{r^2}{R^2}\right)e^{i\omega t} + c.c. \quad (7.39)$$

that matches with the limit of solution (7.36).

From the knowledge of the solution (7.36), the average velocity can be computed through integration over the cross-section of the vessel and dividing by its area

$$U(t) = \frac{2}{R^2}\int_0^R v_x(t,r)r\,dr = \frac{\kappa}{2\omega}\left[1 - \frac{2}{W}\sqrt{\frac{2i}{\pi}}\frac{J_0\left(W\sqrt{\frac{-i\pi}{2}}\right)}{J_1\left(W\sqrt{\frac{-i\pi}{2}}\right)}\right]e^{i\omega t}, \qquad (7.40)$$

where we used the integration formula for the Bessel function $\int u\,J_0(u)du = u\,J_1(u)$. Formula (7.40) represents a sinusoidal function whose amplitude and phase vary with the Womersley number. The amplitude, normalized with $\frac{\kappa}{\omega}$, and the phase of the time function (7.38) are drawn in Fig. 7.7.

For small values of W, the flow is represented by the asymptotic solution (7.39) that, in these units, grows as $\frac{\kappa}{\omega}\frac{\pi}{16}W^2$ and has no phase difference from the external forcing. Vice versa, for large W viscosity is less relevant and the balance is dominated by the inertial term, thus the amplitude is equal to $\frac{\kappa}{\omega}$ and its phase is in quadrature with the forcing.

(ii)   Pulsatile flows

The oscillatory solution describes above is useful to understand the main phenomena entering into play in general time-period flow. Indeed, the flow in cardiovascular vessels is usually pulsatile: unsteady, periodic in time, with non-zero time-average velocity. In a straight vessel, pulsatile flow is obtained by a combination of steady

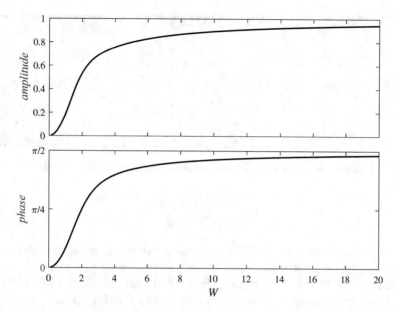

Fig. 7.7 Amplitude and phase of the average velocity in a vessel subjected to sinusoidal pressure gradient

flow and a series of sinusoidal oscillations. Under the hypothesis of unidirectional flow, the transport term in the Navier–Stokes equation is absent and equations are linear (see (7.24) and (7.34)). Therefore, the solution corresponding to an arbitrary time-periodic pressure gradient

$$\frac{1}{\rho}\frac{\partial p}{\partial x} = -\kappa_0 + \sum_n \kappa_n e^{i2\pi n\frac{t}{T}}; \tag{7.41}$$

or an arbitrary mean velocity

$$U(t) = U_0 + \sum_n U_n e^{i2\pi n\frac{t}{T}}; \tag{7.42}$$

can be obtained by appropriate linear combination of solution (7.25) and (7.35), or (7.28) and (7.36).

A Reynolds number can also be introduced for the pulsatile flows as $Re = \frac{UD}{\nu}$, using a velocity scale that can be the mean value or the amplitude of the oscillation, that tells how intense is the bulk flow with respect to the ability of viscous dissipation.

In unsteady periodic flows, it is sometimes useful to introduce another dimensionless number, the Strouhal number (see also Eq. (1.16)), defined as

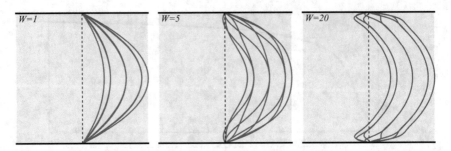

Fig. 7.8 Pulsatile velocity profile in a circular vessel at different Womersley numbers

$$St = \frac{D}{UT} = \frac{W^2}{Re};$$ (7.43)

The Strouhal number represents a dimensionless frequency of the oscillation. It can be appreciated that the length traveled by particles during an oscillation is proportional to UT; therefore, the Strouhal number can be seen as the ratio between the diameter and a measure of the distance traveled by particles during a period. For high St, the oscillations are rapid and fluid particles oscillate for a length smaller than the diameter; low St means that particles travel several diameters during each oscillation.

The solutions for pulsatile flow are then given by a Poiseuille parabolic flow made with the mean velocity plus the individual solutions of sinusoidal flows (like those in Fig. 7.6). Examples of pulsatile flows solutions, given by a mean flow and a single sinusoidal oscillation of amplitude equal to the mean flow, are shown in Fig. 7.8 for the same average velocity and different values of the Womersley number. For low values of W the velocity profile is a sequence of Poiseuille solutions evaluated with the instantaneous values of the mean velocity; on the opposite end, as W increases, the solution presents an inversion of the boundary layer flow in the annulus near the wall.

To give an idea of the order of magnitude for these dimensionless numbers in the circulation, the following table provides indications of typical values in main vessels. It shows that flow is effectively unsteady in major arteries of clinical interest while it becomes well described by the Poiseuille solution in smaller vessels as well as in veins.

	U [cm/s]	D [cm]	Re	W	St
Aorta	100	3	10,000	20	0.04
Middle arteries	30	1	1000	5	0.05
Small arteries	5	0.2	30	1	0.05
Arterioles	0.1	<0.1	<0.5	<0.5	~1

Chapter 8
Elements of Turbulent Flow

Abstract The unidirectional flow becomes unstable as soon as the Reynolds number increases above a certain threshold; beyond which flow enters in a turbulent regime. Turbulence is characterized by a complex three-dimensional motion that, despite it is governed by deterministic laws (Navier–Stokes equation), displays apparently random properties; the unpredictability of deterministic laws is explored through a parallel with simple chaotic systems. The randomness of individual solution suggests to seek a description of turbulent flow in terms of average statistical properties. The classic averaging operator is introduced and applied to the Navier–Stokes equation to obtain an equation for the average velocity, the Reynolds equation. However, the simplification introduced by the well-behaving average variables is paid by the appearance of novel terms in the averaged equation that is non-closed and requires additional assumptions to be solved. The Reynolds equation is used to obtain the general solution for a fluid moving in a turbulent regime over a wall, the result is then used to estimate the friction associated with turbulent flowing inside a duct. Finally, the concept of phase-averaging for time-periodic flows is introduced.

8.1   Introduction to Turbulence

We briefly anticipated that the Reynolds number represents the ratio between the available kinetic energy that corresponds to the thrust of the flow $\Delta p \sim \rho U^2$, and the viscous stress that represents the ability to dissipate kinetic energy by regular friction $\tau \sim \mu \frac{U}{D}$. When the Reynolds number exceeds a certain threshold, the kinetic energy can be excessively high, or the ability to dissipate can be insufficient, and the regular motion is unable to reach a balance between incoming energy and dissipation. In that case, the regular flow solution, called "laminar flow" because assumes fluid streaming across parallel laminae, is unstable and the fluid develops turbulent swirling motion to dissipate the excess energy. This point was analyzed systematically for the first time in the famous experiment performed by Osborne Reynolds (back in 1883) describing the transition from laminar to turbulent flow in a circular pipe under steady and uniform conditions (Reynolds 1894, 1883).

© The Author(s), under exclusive license to Springer Nature Switzerland AG 2022   115
G. Pedrizzetti, *Fluid Mechanics for Cardiovascular Engineering*,
https://doi.org/10.1007/978-3-030-85943-5_8

In that experiment, water was allowed to flow in a glass-walled (transparent) pipe and a small amount of dye was released continuously at the center of the pipe near the inlet. When fluid velocity was small enough, the dye trajectory was rectilinear. This type of motion was said to be laminar; fluid motion is unidirectional and uniform in agreement with the hypothesis used for the Poiseuille solution; thus, the velocity field was described by solution (7.25). When velocity approaches a certain critical threshold, the dye trajectories started to display a slightly wavy pattern indicating that the flow was not perfectly laminar. As velocity further increased above a critical value, the dye rapidly mixes and spreads filling the entire pipe. This type of flow was said "turbulent"; velocities are irregular in space and in time with an apparently random behavior.

Application of dimensional analysis to such an experiment tells that any kinematic flow property (like, for example, the amplitude of turbulent fluctuations) must depend on the size of the circular pipe, measured by its diameter D, by the intensity of the flow, given by the mean velocity U, and by the properties of the flow, which in Newtonian flow are summarized by the kinematic viscosity ν. There were no other parameters that could be varied in the experiment. These are 3 parameters whose dimensions are described by 2 units (length and time); from dimensional analysis, any properly normalized dimensionless quantity can depend on a single dimensionless parameter constructed using a proper combination of such dimensional quantities. The dimensionless parameter is the Reynolds number (this is where the name comes from) $Re = \frac{UD}{\nu}$. When the Reynolds number is below a critical value Re_{cr} the flow is laminar, when it is above the critical value the flow is turbulent. The transition from laminar to turbulence occurs over a small interval of about $Re_{cr} \approx 2500$, although the exact figure depends on the degree of disturbance that is present in the experiment.

The same concept applies to other types of flow. For most flow arrangements, it is possible to define a Reynolds number with an appropriate velocity scale U and an appropriate length scale L, e.g., $Re = \frac{UL}{\nu}$, such that flow is laminar when the Reynolds number is below a critical level, whose precise value depends on the specific arrangement, and becomes turbulence when it exceeds it. For flow behind a cylindrical obstacle, these scales are evidently the upstream velocity and the cylinder diameter. In a steady boundary layer developing over a flat plate, as discussed in Sect. 7.1, the only external length scale available is the distance from the origin of the plate x, thus the effective Reynolds number therein, $Re = \frac{Ux}{\nu}$, grows downstream and indicates that the laminar boundary layer remains stable only up to a certain distance from its origin and it becomes turbulent afterward.

For many decades, the physical origin of turbulence was unclear. The laminar Poiseuille flow is a solution of the Navier–Stokes equation, thus, the observation that this solution does not realize may rise doubts on the validity of those equations for describing fluid motion in general. However, this is not the case. The Navier–Stokes equation is a nonlinear equation; as such, it has not necessarily a unique solution and can admit multiple solutions to the same problem. In the case of pipe flow, there is one laminar solution, that is steady uniform and unidirectional, and there are infinitely other unsteady and irregular turbulent solutions. When the Reynolds number is small enough, the viscous friction is large enough to damp the turbulent solution; thus, any irregularities will decay and the solution converges toward the laminar solution that is the only stable and physically realizable solution. Vice versa, when the Reynolds number is larger, the laminar solution becomes unstable because the flow has an excess of energy that cannot be dissipated through the simple linear friction, thus this solution does not realize physically. Any small disturbance to that solution tends to move the flow away from it, whereas a turbulent regime is a stable and realizable configuration. The selection of one or another of the many possible turbulent solutions depends on the details of the initial and boundary conditions. The solution can also jump from one turbulent solution to another when disturbed by small perturbations that are unavoidably present in physical systems.

The point that deterministic equations like Navier–Stokes can give rise to apparently random solutions has been debated for years in the past decades; however, it was recognized that even much simpler deterministic nonlinear equations could present analogous behavior (Davidson 2004). This was described as the concept of deterministic "chaos", where the solution of deterministic equations produces apparently random behaviors (Feigenbaum 1978). Probably the simplest equation exhibiting such behavior is a difference equation (where time-derivative is replaced by a difference between discrete time instants) with no spatial dependence. Thus consider the variable u_k, with values in the open interval $0 < u_k < 1$, where k is the discrete time variable, obeying the following evolution equation (May 1976):

$$u_{k+1} = \lambda u_k(1 - u_k); \tag{8.1}$$

and λ is a parameter that plays a role analogous to the Reynolds number for this abstract evolution equation. Equation (8.1) has a steady solution, immediately found by setting $u_{k+1} = u_k$, which is $u_k = \frac{\lambda - 1}{\lambda}$ (the other solution $u_k = 0$ is out of the interval of definition). Until $\lambda < 3$, this solution is stable, any initial condition eventually converges to the steady solution. When λ increases the steady solution becomes unstable, the system becomes unsteady, for small increases of λ the solution jumps alternatively between two values, as λ increases a little further the solution jumps between more values, until for $\lambda = 4$ the solution oscillates randomly over the entire interval $(0, 1)$. It is important to underline that Eq. (8.1) is deterministic (as the Navier-Stoke equation is); therefore, once the initial condition is set then the associate solution follows univocally, while different realizations are the result of setting different initial conditions. It should be remarked, however, that the solution at a certain time k is extremely sensitive to the value used as an initial condition; extremely close initial conditions rapidly diverge and after a transient period give rise to macroscopically different solutions. Initial differences can be limited to a far decimal digit, or the last significant digit in a numerical calculation, the smaller the difference the longer the transient, eventually the solutions diverge exponentially and become uncorrelated. This behavior is known as "Sensitivity to Initial Conditions" (SIC). Indeed, the smooth dependence of the solution on initial/boundary conditions (that solutions relative to nearby conditions produce nearby solutions) is generally valid for linear systems only and does not necessarily apply to nonlinear systems.

This behavior also applies to the Navier–Stokes equation, with the additional complexity that irregularities occur both over time and over the three spatial dimensions. This means that any small difference in initial or on boundary conditions can give rise to different turbulent solutions. Moreover, any small external perturbation is analogous to small changes to the initial condition for the following evolution and may drive the system across different turbulent solutions.

This means that an experiment about fluid turbulence performed under identical conditions produces different solutions because a laboratory cannot reproduce conditions that are exactly identical to arbitrary accuracy. This is a fundamental conceptual problem that cannot be solved in principle, and in practice with numerical computation methods as well, because the boundary or initial conditions can be known with a finite accuracy only and such small uncertainty reflects large differences in the solution. Moreover, numerical calculus uses a finite accuracy, that may be even different between different computers, and it can happen that numerical solutions of the same determinist equations produce different results when performed on different computers/compilers. As a result, when talking about turbulent flows one cannot focus on an individual realization of that field, that is just one solution among infinitely many others equally possible, and should rather pay attention to the main properties that are common to all turbulent realizations of that flow.

Physically, turbulence enhances energy dissipation and therefore it is normally a threat of excessive energy consumption in the vascular circulation. Another property is the unpredictability of its chaotic fluctuations that makes turbulent flows difficult to control, model, and manage. On the other side, turbulence has several positive implications; first of all, it makes life possible by enhancing mixing and diffusion. While viscous diffusion is an extremely efficient mechanism to distribute substances at very small scales, turbulent dispersion dominates mixing at larger scales. For example, viscous diffusion length, that grows proportionally to $\sqrt{\nu t}$, in water takes a few hundredths of a second to reach one millimeter, a few seconds for one centimeter, and over one hour for one meter. On the contrary, it is in everyone's experience that accelerated turbulent dispersion dominates the mixing and heat propagation at scales larger than, typically, a few millimeters. It is evident how turbulence is ubiquitous in nature and how it ensures the mixing that is experienced in everyday life.

8.2   Average Fields and Reynolds Equations

Turbulent flows are complex and irreproducible; nevertheless, the different realizations of turbulent flows under similar conditions present common characteristics, like the mean velocity or the amplitude of fluctuations. These properties are also those that present a practical interest. The most common strategy to tackle the problem of turbulence relies on statistical methods, searching for a description of the average

motion (responsible for transport) and of the statistical properties of turbulent fluctuations (responsible for dispersion). This is such a common practice that the study of
turbulence is often considered that of statistical fluid mechanics (Monin and Yaglom
1971).

Indicating with an overbar the averaging operator, a turbulent function, e.g., the
velocity field v, can be expressed as the sum of its average value $\bar{v}$ and the fluctuation
v'

$$v(t, x) = \bar{v}(t, x) + v'(t, x). \tag{8.2}$$

which implies that $\bar{v'} = 0$.

There are several different ways for defining an average operator in turbulence,
from spatial to time averaging, to filtering, with different filtering functions acting
over a finite region of space and/or time. Before choosing a specific averaging operator among the infinite possibilities, it is important to define the properties that it
must satisfy. In the classical approach to turbulence, the Reynolds average operator
is used, which restricts the choice of those operators that satisfy the Reynolds rules
(Monin and Yaglom 1971). These comprise the linearity conditions that given two
fields u and v and two constants a and b

$$\overline{au + bv} = a\bar{u} + b\bar{v}; \tag{8.3}$$

the commutation with time or space derivative

$$\overline{\frac{\partial v}{\partial t}} = \frac{\partial \bar{v}}{\partial t}, \ \overline{\frac{\partial v}{\partial x}} = \frac{\partial \bar{v}}{\partial x}; \tag{8.4}$$

and the invariance

$$\overline{\bar{u}v} = \bar{u}\bar{v}, \tag{8.5}$$

which can be intuitively understood considering that the average term behaves like a
constant with respect to the averaging operation because it does not vary within the
(space or time) domain over which the averaging is performed.

Reynolds used the most straightforward approach to averaging, considering the
time average over a period T that defines a duration longer than of fluctuations due to
turbulence under the assumption that the time changes associated with the evolution
of average flow occur over much longer time scales

$$\bar{v}(t, x) = \frac{1}{T} \int_{-\frac{T}{2}}^{+\frac{T}{2}} v(t + t', x) dt'; \tag{8.6}$$

We are now ready to write the equations for the mean velocity applying the Reynolds average operator to the continuity and the Navier–Stokes equations. Applying the mean operator to the continuity equation and using (8.4) gives

$$\overline{\nabla \cdot \boldsymbol{v}} = \nabla \cdot \overline{\boldsymbol{v}} = 0; \tag{8.7}$$

which tells that the mean velocity is also a divergence-free field. By difference, it is immediate to verify that also the fluctuating velocity field is divergence-free

$$\nabla \cdot \boldsymbol{v} = \nabla \cdot \left(\overline{\boldsymbol{v}} + \boldsymbol{v}' \right) = \nabla \cdot \overline{\boldsymbol{v}} + \nabla \cdot \boldsymbol{v}' = \nabla \cdot \boldsymbol{v}' = 0. \tag{8.8}$$

The same approach can be applied to the Navier–Stokes equation. To make it simple, consider the x-component of the equation written in a Cartesian system of coordinates, see Eq. (5.34), that when averaged gives

$$\frac{\partial \overline{v}_x}{\partial t} + \overline{\boldsymbol{v} \cdot \nabla v_x} = -\frac{1}{\rho}\frac{\partial \overline{p}}{\partial x} + \nu \nabla^2 \overline{v}_x. \tag{8.9}$$

where we used rules (8.3) and (8.4) to separate the terms and exchange derivative and averaging operator. The second term in (8.9) is nonlinear and does not allow a simplification based on the commutation (8.4) between averages and derivatives. Let us look at this term in detail

$$\overline{\boldsymbol{v} \cdot \nabla v_x} =$$

$$= \overline{\left(\overline{v}_x + v'_x \right) \frac{\partial \left(\overline{v}_x + v'_x \right)}{\partial x} + \left(\overline{v}_y + v'_y \right) \frac{\partial \left(\overline{v}_x + v'_x \right)}{\partial y} + \left(\overline{v}_z + v'_z \right) \frac{\partial \left(\overline{v}_x + v'_x \right)}{\partial z}}$$

$$= \overline{\overline{v}_x \frac{\partial \overline{v}_x}{\partial x} + \overline{v}_y \frac{\partial \overline{v}_x}{\partial y} + \overline{v}_z \frac{\partial \overline{v}_x}{\partial z} + v'_x \frac{\partial v'_x}{\partial x} + v'_y \frac{\partial v'_x}{\partial y} + v'_z \frac{\partial v'_x}{\partial z}}$$

$$\overline{+ \overline{v}_x \frac{\partial v'_x}{\partial x} + \overline{v}_y \frac{\partial v'_x}{\partial y} + \overline{v}_z \frac{\partial v'_x}{\partial z} + v'_x \frac{\partial \overline{v}_x}{\partial x} + v'_y \frac{\partial \overline{v}_x}{\partial y} + v'_z \frac{\partial \overline{v}_x}{\partial z}}.$$

These terms can be simplified further by realizing that the average values can be taken out of the average operator based on the invariance rule (8.5). Therefore, we can rewrite the last equality

$$\overline{\boldsymbol{v} \cdot \nabla v_x} =$$

$$= \overline{v}_x \frac{\partial \overline{v}_x}{\partial x} + \overline{v}_y \frac{\partial \overline{v}_x}{\partial y} + \overline{v}_z \frac{\partial \overline{v}_x}{\partial z} + \overline{v'_x \frac{\partial v'_x}{\partial x}} + \overline{v'_y \frac{\partial v'_x}{\partial y}} + \overline{v'_z \frac{\partial v'_x}{\partial z}}$$

$$+ \overline{v}_x \frac{\partial \overline{v'_x}}{\partial x} + \overline{v}_y \frac{\partial \overline{v'_x}}{\partial y} + \overline{v}_z \frac{\partial \overline{v'_x}}{\partial z} + \overline{v'_x} \frac{\partial \overline{v}_x}{\partial x} + \overline{v'_y} \frac{\partial \overline{v}_x}{\partial y} + \overline{v'_z} \frac{\partial \overline{v}_x}{\partial z}$$

It is immediate to notice that the first group of three term on the right-hand side is the transport equation written for the mean velocity. The third and fourth groups (on the second line) are both zeroes because they contain the mean of the fluctuating components that are zero by definition. The second group of three terms can be simplified by adding the average of a null term $v_x \nabla \cdot v'$ that is zero because of (8.8); thus, the equality can be rewritten as

$$\overline{v \cdot \nabla v_x} = \bar{v} \cdot \nabla \bar{v}_x + \overline{v'_x \frac{\partial v'_x}{\partial x}} + \overline{v'_y \frac{\partial v'_x}{\partial y}} + \overline{v'_z \frac{\partial v'_x}{\partial z}} + \overline{v'_x \frac{\partial v'_x}{\partial x}} + \overline{v'_x \frac{\partial v'_y}{\partial y}} + \overline{v'_x \frac{\partial v'_z}{\partial z}}$$

$$= \bar{v} \cdot \nabla \bar{v}_x + \frac{\partial \overline{v'_x v'_x}}{\partial x} + \frac{\partial \overline{v'_x v'_y}}{\partial y} + \frac{\partial \overline{v'_x v'_z}}{\partial z};$$

where we used the product of derivatives in the last passage.

This result can be reinserted in the original Eq. (8.9) and rewritten in vector terms to better highlight the new structure of the equation

$$\frac{\partial \bar{v}}{\partial t} + \bar{v} \cdot \nabla \bar{v} = -\frac{1}{\rho} \nabla \bar{p} + v \nabla^2 \bar{v} - \frac{1}{\rho} \nabla \cdot \mathbb{T}_R, \tag{8.10}$$

where the last term is written using the symmetric Reynolds stress tensor defined by

$$\mathbb{T}_{Rij} = \rho \overline{v'_i v'_j}. \tag{8.11}$$

Equation (8.10) is the Reynolds equation. It corresponds to the Navier–Stokes equation when expressed in terms of mean velocity. The Reynolds equation differs from the Navier–Stokes equation for the additional terms that contain the Reynolds stress. Interestingly, the Reynolds stress term has exactly the same form as the stress tensor term previously found in the Cauchy Eq. (5.24); however, this equation is for the mean velocity, which is not the physical velocity but only a mathematical filter of it. The Reynolds stresses, therefore, are not real stresses experienced by physical fluid elements, they are fictitious stresses that represent the influence of the turbulent fluctuations on the mean velocity. The Reynolds stress tensor accounts for the energy that is lost by the mean flow because it is transferred into the turbulent fluctuations.

The Reynolds equation produces simpler, smoother solutions because of the enhanced dissipative mechanism introduced by the Reynolds stresses, thus avoiding the contemporary presence of many interleaving scales within the flow. This simplification is payed, on the other side, by the fact that the Reynolds equation is not closed: it includes 6 further unknowns, the Reynolds stresses, which cannot be obtained by the equation itself. The appearance of novel unknowns in the averaged equation is what is known as the *closure problem* of turbulence. Either equations are complicated and unsolvable (Navier–Stokes) o they are not closed (Reynolds) because they present additional unknown terms.

There are numerous models to provide a closure to the Reynolds equation by adding additional equations for the terms (8.11). It must be remarked, however, that all such models are not obtained from first conservation principles; thus, closure models are approximations, they can be non-accurate in general and rely on numerous empirical coefficients. They are more reliable in canonical flows of practical relevance where extensive experimental and numerical studies permitted to estimate such coefficients and establish reliable models. In general, the closure problem is still open, although several advances have been performed during the past decades to allow numerical approximate solutions of turbulent flows (Sagaut 2006).

8.3 Turbulent Flow Over a Wall

A turbulent flow of paramount applied interest is that flowing near a solid surface under the hypothesis that the mean flow is steady and unidirectional. It cannot be solved exactly, as it was done for laminar flows, because of the closure problem. Nevertheless, some results can be achieved by properly combining all information available.

Following (Monin and Yaglom 1971), consider a turbulent flow over a flat surface as sketched in Fig. 8.1. Assume that the average flow is steady and two dimensional, thus derivatives of mean quantities along t and z, assumed the transversal direction, are zero. Also, assume that the mean flow is unidirectional, $\bar{v}_y = \bar{v}_z = 0$. By continuity, we have only unknown is the x-component of the mean velocity that can vary with the distance y from the wall: $\bar{v}_x(y)$.

The Reynolds equation along the transversal direction simply states that the mean pressure $\bar{p}$ is constant transversal to the mean flow, as can be immediately verified. The x-component of the Reynolds equation

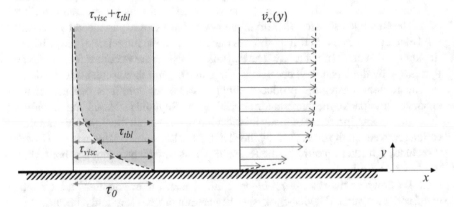

Fig. 8.1 Turbulent flow over a flat surface

$$\frac{\partial \bar{v}_x}{\partial t} + \bar{v}_x \cdot \frac{\partial \bar{v}_x}{\partial x} + \bar{v}_y \cdot \frac{\partial \bar{v}_x}{\partial y} + \bar{v}_z \cdot \frac{\partial \bar{v}_x}{\partial z} =$$

$$= -\frac{1}{\rho}\frac{\partial \bar{p}}{\partial x} + \nu\left(\frac{\partial^2 \bar{v}_x}{\partial x^2} + \frac{\partial^2 \bar{v}_x}{\partial y^2} + \frac{\partial^2 \bar{v}_x}{\partial z^2}\right) - \frac{\partial \overline{v'_x v'_x}}{\partial x} - \frac{\partial \overline{v'_x v'_y}}{\partial y} - \frac{\partial \overline{v'_x v'_z}}{\partial z},$$

simplifies with these assumptions to

$$\nu\frac{\partial^2 \bar{v}_x}{\partial y^2} - \frac{\partial \overline{v'_x v'_y}}{\partial y} = \frac{1}{\rho}\frac{\partial \bar{p}}{\partial x}.$$

Make the additional assumption that the mean pressure gradient is zero; such that the motion is driven by the presence of a velocity away from the wall. However, it can be demonstrated that the presence of a mean pressure gradient, which is constant over y, would affect the value of velocity away from the wall but without a direct influence on the profile of velocity (Monin and Yaglom 1971). In this simple formulation, the Reynolds equation provides the following relationship

$$\nu\frac{\partial^2 \bar{v}_x}{\partial y^2} - \frac{\partial \overline{v'_x v'_y}}{\partial y} = 0. \tag{8.12}$$

It is noticed that both terms in (8.12) present a derivative along y; after integration we have

$$\nu\frac{\partial \bar{v}_x}{\partial y} - \overline{v'_x v'_y} = constant;$$

this constant has the dimension of a velocity square and it is commonly written as u_*^2

$$\nu\frac{\partial \bar{v}_x}{\partial y} - \overline{v'_x v'_y} = u_*^2, \tag{8.13}$$

where u_* is called the *friction velocity* whose physical meaning will be clear shortly.

Equation (8.13) tells that there is a property of the turbulent wall flow, that we call friction velocity, which is constant over the entire velocity profile, from the wall to above the wall. Looking carefully at the two terms in (8.13) we can recognize that, when multiplied with the density ρ, they represent the mean viscous stress $\mathbb{T}_{xy} = \mu\frac{\partial \bar{v}_x}{\partial y}$ and the turbulent Reynolds stress $\mathbb{T}_{Rxy} = -\rho\overline{v'_x v'_y}$, respectively. Therefore, Eq. (8.13) tells that the total shear stress, given by the sum of viscous plus turbulent stresses, is constant over the turbulent profile; the former dominates close to the wall, where turbulence stress reduces approaching the wall where velocities are zero, while the turbulent stress dominates away from the wall where turbulent fluctuations increase and the influence of viscosity progressively decreases, as shown in Fig. 8.1.

Equation (8.13) is valid also at the wall where turbulent stress is zero because velocity
and its fluctuations are zero and the viscous stress at the wall τ_0 is

$$\tau_0 = \rho v \left. \frac{\partial \bar{v}_x}{\partial y} \right|_{y=0} = \rho u_*^2. \tag{8.14}$$

Equation (8.14) tells that the friction velocity is given by

$$u_* = \sqrt{\frac{\tau_0}{\rho}}. \tag{8.15}$$

Given the global picture, consider first the limiting situation very close to the wall.
Here viscous stress dominates and Eq. (8.13) can be approximated as

$$v \frac{\partial \bar{v}_x}{\partial y} \cong u_*^2,$$

which can be integrated, with boundary condition $\bar{v}_x(0) = 0$, to give the velocity
profile very close to the wall showing that velocity grows linearly from the wall

$$\frac{\bar{v}_x}{u_*} \cong \frac{u_*}{v} y. \tag{8.16}$$

Expression (8.16) indicates the existence of a viscous length scale, given by $y_* = \frac{v}{u_*}$, at which distance $\bar{v}_x(y_*) \cong u_*$, that provides the order of magnitude of the size
of this inner layer dominated by viscous stresses.

Far from the wall, at distances much larger than such length scale, $y \gg y_*$, the
turbulent stress dominates over the viscous one. Equation (8.13) does not allow to
solve the velocity profile in this limit; nevertheless, we can confidently assume that
the variations in the velocity profile (its y-derivative) should not depend explicitly
on viscosity. Thus, in a first approximation, the velocity gradient can only depend
on the turbulent stress, whose magnitude is estimated from (8.13) to be given by the
friction velocity, and by the distance from the wall that is a length scale of the spatial
amplitude of turbulent fluctuation. Functionally we can write

$$\frac{\partial \bar{v}_x}{\partial y} = f(u_*, y). \tag{8.17}$$

The relation (8.17) represents a physical law and, even though the function f
is unknown, is must be dimensionally consistent. The dimensional analysis allows
simplifying it in the following form:

$$\frac{\partial \bar{v}_x}{\partial y} = \frac{1}{k} \frac{u_*}{y}; \tag{8.18}$$

where k is a constant that must be evaluated experimentally. This constant is known as the Von Karman constant, it was estimated to take the value $k \cong 0.4$ in most turbulent wall flows. Integration of (8.18) gives

$$\overline{v}_x = \frac{u_*}{k} \ln\left(\frac{y}{y_0}\right);$$
(8.19)

where y_0 is the integration constant that is unknown because there is no boundary condition that can be enforces given that this profile is not valid at the wall. The integration constant y_0 has the dimension of a length and should be expressed proportional to the only existing length scale y_*, and can be written as $y_0 = \frac{1}{a}\frac{\nu}{u_*}$ transferring the lack of knowledge from y_0 to the dimensionless constant a. With this substitution, Eq. (8.19) is rewritten in a form analogous to (8.16) as

$$\frac{\overline{v}_x}{u_*} = \frac{1}{k} \ln\left(a\frac{u_*}{\nu}y\right),$$
(8.20)

where the coefficient a is unknown and should be evaluated experimentally for the different situations. The solution (8.20), although obtained with several approximations and hypotheses, was demonstrated to be a very good representation of real wall-bounded turbulent flows under numerous different configurations, with or without pressure gradients, and in different geometries. It is valid for the wind blowing over the sea or over a town, for water flowing in rivers as well as in cylindrical pipes. A distinctive feature of profile (8.20), sketched in Fig. 8.1, is its slow modulation after an initial steep region close to the wall. Therefore, the turbulent profiles are commonly very flat in contrast with the laminar parabolic profile.

The velocity profile in Eq. (8.20) is not an exact solution; it presents two dimensionless coefficients, k and a, that must be estimated experimentally, and the friction velocity u_* that provides the intensity of the actual flow. Friction velocity was defined by (8.15) and it is the only velocity scale available in this context of generic flow without reference to conduits discharge or external velocities. However, once the details of the flow are provided, the friction velocity can be obtained from macroscopic measurable quantities. For example, in a conduit, the friction velocity can be related to the pressure gradient that induces the mean flow. To make this point explicit, consider a steady and uniform turbulent flow inside a vessel of constant cross-section, the global balance of momentum between two sections has zero inertial and zero flux of momentum terms, it simply states a balance between the force due to pressure difference and the resistance due to wall shear stress

$$\tau_0 C = -\frac{\partial \overline{p}}{\partial x} A,$$
(8.21)

where C and A are the perimeter and the area of the vessel, respectively; equal to πD and $\frac{\pi}{4}D^2$ in a circular vessel of diameter D. Use (8.15) and (8.21) to obtain the friction velocity in terms of the pressure gradient,

$$u_* = \sqrt{-\frac{1}{\rho}\frac{\partial \overline{p}}{\partial x}\frac{A}{C}} = \sqrt{-\frac{1}{\rho}\frac{\partial \overline{p}}{\partial x}\frac{D}{4}} \qquad (8.22)$$

with the second equality valid for a circular vessel only.

We can also move further and use the previous results to build the relationship between friction velocity, or pressure gradient, and average velocity in a vessel of diameter D as follows. The logarithmic profile is valid over a large portion of the duct. Therefore, there must be a certain distance from the wall y where the local velocity is equal to the average velocity U in the duct. Express this distance as proportional to the duct diameter $y = a'D$, with a' an unknown constant; we can write in formulas that a value a' must exist such that $\overline{v}_x(a'D) = U$. Inserting this condition in (8.20) gives

$$\frac{U}{u_*} = \frac{1}{k}\ln\left(b\frac{UD}{\nu}\frac{u_*}{U}\right), \qquad (8.23)$$

where b is the new unknown constant $(b = aa')$, which was experimentally estimated in circular vessels to be $b \cong 1.13$.

Equation (8.23) provides a relationship between mean velocity and pressure gradient (or friction velocity). To this aim, in turbulence, it is common habit to introduce a friction coefficient, the dimensionless Chezy coefficient, as the ratio between mean and friction velocities, $C = \frac{U}{u_*}$, such that the relationship (8.23) be rewritten as

$$C = \frac{1}{k}\ln\left(b\frac{Re}{C}\right), \qquad (8.24)$$

which is an implicit expression for $C(Re)$. Then extracting the pressure gradient from (8.22) after substituting $u_* = CU$ gives

$$-\frac{dp}{dx} = \frac{8}{C^2(Re)}\frac{\rho U^2}{2D}, \qquad (8.25)$$

where we removed the average operator on the pressure. Expression (8.25) has a form identical to Eq. (7.31) where the friction coefficient $f(Re)$ was substituted by the Chezy coefficient; the two are related by

$$C = \sqrt{\frac{8}{f}}, \quad f = \frac{8}{C^2}. \qquad (8.26)$$

The friction coefficient was evaluated in (7.32) for the Poiseuille flow; thus, the Chezy coefficient in Poiseuille flow is $C = \sqrt{\frac{Re}{8}}$. The two friction coefficients can be used interchangeably, depending on the traditions in different contexts, geographical regions, or disciplines.

The previous evaluations permit to have initial estimates of pressure loss and velocity profile for the simple case of steady, uniform turbulent flows. In unsteady and in spatially non-uniform flows, expressions for the wall shear stress and for the energy losses are not available in general. Few results are in laminar flows (like those in Sect. 7.4) and almost none in turbulent flows.

8.4 Phase-Average and Steady Streaming

Fully developed, statistically steady turbulence is not common in the cardiovascular system. Flow in the largest arteries and in the cardiac chambers shows the presence of irregular behavior that is often classified as weak turbulence and that is due to the complex interaction of vortices, which will be discussed in more depth in Chap. 10. For this reason, as well as for the physiological variability, the flow field during one heartbeat is never identical to another.

The beat-to-beat variability must be accounted for when analyzing cardiovascular data. Actually, some imaging technology, like magnetic resonance, builds images collecting data received during numerous heartbeats and returns results corresponding to a single average heartbeat. However, many other recordings, like echocardiography or electrocardiograms, as well as in in vitro experiments, return data recorded in real time over several heartbeats. In those situations, as well as in general to periodic pulsatile flows, it is often of interest to extract properties that are common to every heartbeat and consider beat-to-beat changes as "fluctuations" imputable to turbulence or to secondary physiological variations. The concept of averaging for periodic pulsatile flows requires a dedicated approach that differs from the classical averaging discussed above.

Assume that the velocity field $v(t, x)$ at one or more spatial positions, x, is recorded over a number N of heartbeats of period equal to T. The *phase-average* velocity is defined as the average made from the various heartbeat at the same relative instant inside the individual heartbeat. In the formulae, the phase-average of the velocity field (still indicated with an overbar) is defined by

$$\bar{v}(t, x) = \frac{1}{N} \sum_{n=0}^{N-1} v(nT + t, x). \qquad (8.27)$$

there $t \in [0, T)$. This formula can be readily modified to account for a period that is (slightly) variable and takes values $T_k, k = 1 \ldots N$, with average value T, substituting $nT + t$ with $\frac{1}{n} \sum_{k=1}^{n} T_k + \frac{T}{T_{k+1}} t$ in the time argument of the recorded velocity on the right-hand side. In such case, however, it is often simpler to stretch/contract the time in every heartbeat and reconduct the analysis to a situation where T is constant and using the correction coefficient $\frac{T}{T_k}$ when performing time derivatives, integrals or analogous operations.

In practice, the process of phase-averaging eliminates the fluctuations due to the background turbulence and or minor beat-to-beat physiological variations and extracts only the component corresponding to a single average period from the flow recorded over several beats. Being the average velocity (8.27) time periodic, with period T, it can also be expressed in Fourier series as

$$\bar{v}(t, x) = v_0(x) + \sum_{k=1}^{\infty} v_k(x) e^{i\frac{2\pi kt}{T}} + c.c., \tag{8.28}$$

where the individual harmonics $v_k(x)$ can be evaluated from the original velocity by

$$v_k(x) = \frac{1}{NT} \int_0^{NT} v(t, x) e^{-i\frac{2\pi kt}{T}} dt. \tag{8.29}$$

The zero harmonic $v_0(x)$

$$v_0(x) = \frac{1}{T} \int_0^T v(t, x) dt. \tag{8.30}$$

represents the time-averaged velocity over the heartbeat that is often referred to as the *steady streaming* component (Riley 2001). Steady streaming represents the average motion of a periodic flow and it is often useful to extract the net result of the unsteady motion into a single time-frozen field clearing out the effect of time modulation.

After the phase-average operation is defined, it is possible to write a decomposition analogous to (8.2)

$$v(t, x) = \bar{v}(t, x) + v'(t, x). \tag{8.31}$$

and derive an equation for the mean velocity following a similar approach to that used above for obtaining the Reynolds equation (Reynolds and Hussain 1972) that, like Reynolds equation, is not closed and requires introducing approximate closure schemes to be solved. However, the usage of those equations is not common in applications, and the phase-averaging, using operator (8.27) or the Fourier representation (8.28)–(8.29), is most commonly applied in post-processing of data recorded from multiple heartbeats.

On the basis of the phase-average flow field, it is possible to evaluate numerous quantities that further characterize the phase-averaged flow. However, it must be underlined that the properties of the average flow do not always correspond to the average property of the flow. The average value is identical, in principle, to the value associated with the average flow only when that property presents a linear relation with velocity. This is the case, for example, of the flow inertia (velocity time derivative) or of the transport of a scalar quantity. Differently, they disagree when the quantity of interest presents a nonlinear relation with velocity because nonlinear operations do not commute with the averaging operators. This is the case, for example, of the kinetic energy that is given by the square of velocity; the kinetic energy evaluated from the average velocity differs from the average value of kinetic energy that includes a contribution coming from fluctuations; other examples include the energy dissipation or the stress tensor (Andersson et al. 2019). In those cases, it is necessary to apply the averaging operator (8.27) directly to each individual quantity, otherwise to specify explicitly that they are properties evaluated from the average velocity field and not average value of those properties.

Chapter 9
Quasi-Unidirectional Flow in Large Vessels

Abstract The cardiovascular system presents a complex geometrical structure that reflects into features of the fluid motion therein. This chapter analyzes some of the main phenomena encountered in main vessels that allow analysis in simple terms and can provide information regarding general properties of the cardiovascular function. It starts with investigating the implications of mass balance on tapering geometry, followed by the development of secondary flow due to the curvature of main arteries. Then the classic result regarding the propagation of elastic waves is presented introducing the concept of celerity related to arterial stiffness. The same subject is extended to the transmission and reflection of elastic waves at a bifurcation, which is analyzed in general and discussed with reference to the aorta. Finally, a mention is given about collapsible vessels and the concept of flow limitation.

9.1 Mass Balance in Tapering and Branching Arteries

Tapered geometry, where the cross-sectional area decreases downstream, and branches, extracting flow from the main vessel, are characteristic elements in many arteries. Let us verify what mass balance tells about these situations. Consider first a tapered vessel and assume the duct as non-deformable. The discharge $Q = UA$ is constant along the vessel, thus when area A decreases velocity U must increase following mass balance equation

$$\frac{dQ}{dx} = A\frac{dU}{dx} + U\frac{dA}{dx} = 0; \tag{9.1}$$

from which the velocity should increase at a rate

$$\frac{dU}{dx} = -\frac{Q}{A^2}\frac{dA}{dx} > 0. \tag{9.2}$$

This result does not realize physiologically because velocity must decrease when the vessel size decreases to avoid excessive friction that would result from a relationship like (7.29).

Indeed, in real arteries, the discharge decreases downstream $\frac{dQ}{dx} < 0$ in virtue of the side branches. At the same time, the velocity must decrease downstream $\frac{dU}{dx} < 0$ to avoid an increase of friction; these considerations give a relationship between area reduction and discharge reduction. Extract the velocity gradient from mass balance (9.1) and impose that, differently from (9.2) it must be negative

$$\frac{dU}{dx} = \frac{1}{A}\frac{dQ}{dx} - \frac{Q}{A^2}\frac{dA}{dx} < 0. \tag{9.3}$$

Condition (9.3) can be restated as

$$-\frac{1}{Q}\frac{dQ}{dx} > -\frac{1}{A}\frac{dA}{dx}; \tag{9.4}$$

telling that the relative (percentage) reduction of discharge must be larger than the relative reduction of area, otherwise, velocity would increase downstream.

This argument becomes more immediate when applied to bifurcations. Consider a vessel with an area A_0 and discharge $Q_0 = U_0 A_0$, where U_0 is the velocity, that bifurcates into two equal daughter vessels, each of area A_1 and discharge $Q_1 = U_1 A_1$. Mass conservation tells

$$U_0 A_0 = 2 U_1 A_1, \tag{9.5}$$

where, as discussed before, we want the condition that velocity reduces in smaller vessels, $U_1 < U_0$ when $A_1 < A_0$. Using (9.5) this implies that

$$\frac{U_0}{U_1} = \frac{2A_1}{A_0} > 1. \tag{9.6}$$

Thus, although the individual daughter vessels reduce their size, the sum of their areas must increase. Indeed, the total cross-sectional area increases downstream at every branching.

To get an idea of this geometric effect, consider the diameter of the aorta, the first artery after the heart, whose diameter is approximately 3 cm. The total cross-section of the blood vessel at the root of the aorta is approximately $A_{Aorta} \approx 7\,\mathrm{cm}^2$ where the flow has a velocity of about $U_{Aorta} \approx 1\,\mathrm{m/s}$, which corresponds to a discharge $Q \approx 700\,\mathrm{cm}^3/\mathrm{s}$. A similar discharge must cross the entire cross-section of the vasculature at any level of branching. Consider that at the capillary level blood velocity is smaller than 1 mm/s, this means that the total cross-section of the capillary bed is close to 1 m². Thus, the cross-sectional area increases over 1000 times during progressive branching.

9.2 Flow in Curved Vessels

The motion of a fluid in a curved in curved vessel presents some differences with respect to the laminar flow in a straight vessel (Poiseuille flow), because fluid particles cannot proceed by parallel trajectories given that particles on the inner side of the bend would travel a shorter path than those on the external side.

From a dynamic perspective, fluid particles are subjected to centrifugal acceleration, proportional to the square of their local velocity and inversely proportional to the curvature of the trajectory, v^2/R. Particles on the internal side have a smaller radius of curvature and those near the center of the vessel have a higher velocity. As a result, a pressure gradient develops transversally to the main flow direction pushing toward the external side across the center of the vessel. Then, for conservation of mass, flow returns from the external side to the internal side along the walls. This gives rise to circulatory patterns on planes transversal to the main flow, as sketched in Fig. 9.1, which are called "secondary circulations" (Pedley 1980).

Flow in curved vessels always develops secondary circulations. These take the form of two symmetric circulating cells when the curvature is planar, when the system presents a mirror symmetry relative to the plane containing the curve. Most arteries, however, present a double curvature, mathematically described as curvature and torsion like a portion of a helical duct. A helical curve cannot be contained in a plane; it presents a radius of curvature on a plane and a torsion that depart from that plane. For example, a helix developing along the z direction can be described by its coordinates $x = R\cos\vartheta$, $y = R\sin\vartheta$, $z = c\vartheta$, along a generic parametric coordinate ϑ. When $c = 0$, the curve is planar and can be approximated locally by a portion of a circumference of radius R (curvature $\frac{1}{R}$ and no torsion). When $c \neq 0$, the curve cannot lay on a plane because it presents a torsion and can be approximated locally by a helix with curvature $\frac{R}{R^2+c^2}$, and torsion $\frac{c}{R^2+c^2}$. In presence of a torsion, the symmetry of the two cells is broken of an amount that depends on the degree of torsion. One secondary circulatory cell becomes dominant and occupies the core of the vessel

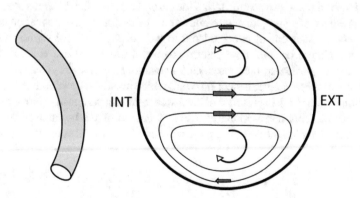

Fig. 9.1 Flow in a curved vessel

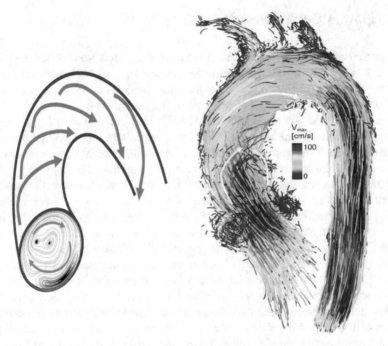

Fig. 9.2 Helical trajectories in the aortic arch. Secondary flow from numerical study and associated sketch (left), measurement from magnetic resonance (right) (credit, right picture: Oechtering et al. J Thorac Cardiovasc Surg 2020;159:798, with permission from Elsevier)

while the other is pushed close to the wall (see in Fig. 9.2, left picture). Flow in real, doubly curved, arteries is thus composed of the main streamwise motion plus a rotation, due to the dominance of one secondary circulation cell. The result of such a combination is that fluid particles move downstream along helical trajectories. This is remarkably noticeable in the aortic arch like that shows in Fig. 9.2. Helical trajectories develop also in many bifurcations, like the carotid and the iliac bifurcations.

The presence of a non-planar curvature and the development of helical trajectories during blood advancement is considered to have a physiological significance and represent a natural optimal flow pattern in several respects (Caro et al. 1996; Liu et al. 2015). In particular, the presence of a small degree of swirl gives rise to open trajectories of blood elements near the wall, which avoid the development of closed stagnation regions even in presence of local area enlargement/constriction or of lateral inflow/outflow. This corresponds to a higher degree of wash-out over the vessel endothelium reducing the risk of aggregation and of development of arteriosclerosis.

9.3 Flow in Elastic Vessels

Arteries are elastic and deform in virtue of the pressure changes associated with the flowing blood inside the vessel. In order to analyze the interaction between vessel elasticity and fluid flow let us first make a premise with a very basic background of solid mechanics that is required to analyze how the deformation occurs in presence of a change in pressure.

Consider a vessel or diameter D and thickness s, assumed small, subjected to an increase of pressure dp inside its lumen. Vessel deformation obeys the law of motion for the elastic material, which simplifies into an equilibrium of forces given that the inertial and transport parts are usually negligible. With reference to Fig. 9.3, internal stresses into the tissue are indicated by τ and they are assumed constant over the thickness. The equilibrium equation for this case is

$$\tau 2s = dpD. \tag{9.7}$$

Equilibrium (9.7) is evaluated in the undeformed configuration, thus assuming that deformations are small (rigorously speaking, infinitesimal); indeed, arterial deformations are usually less than 10% and this approximation is acceptable in this context.

The equilibrium of forces must be combined with the constitutive equation describing the solid material and how internal stresses develop when the tissue deforms. The constitutive equation, for an elastic continuous material subjected to small one-dimensional deformations, is a linear relationship between stress and strain (deformation), which in this case reads

$$\tau = E\frac{dD}{D}, \tag{9.8}$$

where E is the Young modulus of elasticity that describes the elastic stiffness of the tissue. Combination of (9.7) and (9.8) permits to evaluate the change in vessel diameter, dD, in response to an increase of pressure, dp,

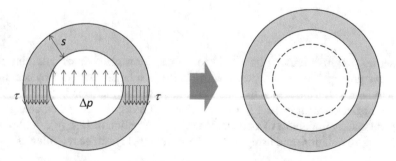

Fig. 9.3 Deformation of an elastic vessel

$$\frac{dD}{D} = \frac{dpD}{2Es}.$$ (9.9)

Then, mass conservation permits to evaluate the change of thickness, that for small thickness gives

$$\frac{ds}{s} = -\frac{dD}{D}.$$ (9.10)

This elementary background on solid mechanics can be integrated with the equation of fluid flow in a vessel to analyze the interaction between fluid flow and elastic tissue. The foundation of the phenomenon of wave propagation in elastic vessels is presented below.

Consider the equation of continuity (4.9) and of motion (5.17) for a vessel

$$\begin{cases} \frac{\partial A}{\partial t} + U\frac{\partial A}{\partial x} + A\frac{\partial U}{\partial x} = 0, \\ \frac{\partial U}{\partial t} + U\frac{\partial U}{\partial x} + \frac{1}{\rho}\frac{\partial p}{\partial x} = 0. \end{cases}$$ (9.11)

In writing the second of (9.11), we made the additional assumption that friction is negligible, as it does not alter qualitatively the propagation phenomenon and would only produce an attenuation of the propagating wave.

The additional relationship needed here is the coupling between vessel size, A, and fluid pressure, p. Making the additional assumption that the vessel has uniform properties along its axis, this relationship does not vary along at different positions and can be expressed as $A(p)$. The existence of a relationship $A(p)$ permits rewriting any derivative of the vessel area, in time or space, in terms of derivative of pressure

$$\frac{\partial A}{\partial x, t} = \frac{dA}{dp}\frac{\partial p}{\partial x, t},$$ (9.12)

where the function $\frac{dA}{dp}$ characterizes the vessel elastic response. For example, in the case of infinitesimal deformation of the linearly elastic vessel discussed above, this function is obtained by the relationship (9.9) that can be recast as

$$\frac{dA}{dp} = \frac{AD}{Es}.$$ (9.13)

Let us simplify further Eqs. (9.11) for propagation phenomena assuming that the wave propagation velocity, the celerity c, is much larger than the physical fluid velocity, $c \gg U$. This means that the second (convective) term in both Eqs. (9.11) can be neglected with respect to the first (inertial) term; indeed, when changes in time are mainly imputable to a phenomenon of propagation then $\frac{d}{dt} \approx c\frac{d}{dx} \gg U\frac{d}{dx}$. In this approximation, and using (9.12), the system (9.11) can be rewritten as

$$\begin{cases} \frac{1}{A}\frac{dA}{dp}\frac{\partial p}{\partial t} + \frac{\partial U}{\partial x} = 0, \\ \frac{\partial U}{\partial t} + \frac{1}{\rho}\frac{\partial p}{\partial x} = 0. \end{cases} \tag{9.14}$$

Now take the time derivative of the former and the space derivative of the latter, and subtract the two results; in doing so, make the additional assumption that the term $\frac{1}{A}\frac{dA}{dp}$ is a property of the vessel that can be considered as slowly and little varying such that we can neglect its derivative. We obtain the equation for pressure

$$\frac{1}{A}\frac{dA}{dp}\frac{\partial^2 p}{\partial t^2} - \frac{1}{\rho}\frac{\partial^2 p}{\partial x^2} = 0.$$

This can be rewritten in canonical form as

$$\frac{\partial^2 p}{\partial t^2} - c^2\frac{\partial^2 p}{\partial x^2} = 0, \tag{9.15}$$

which is the well-known wave equation where

$$c = \sqrt{\frac{A}{\rho}\frac{dp}{dA}} \cong \sqrt{\frac{Es}{\rho D}} \tag{9.16}$$

is the celerity of the wave; the second equality in (9.16) is obtained using Eq. (9.13) and it is valid in the limit case of infinitesimal deformation.

The general solution of the wave Eq. (9.15) is

$$p(t, x) = p(x \pm ct), \tag{9.17}$$

which corresponds to a rigid propagation of the pressure fields without a change of shape.

The previous analysis was outlined in terms of pressure; however, it can be replicated in terms of velocity by subtracting the derivative in space of the former equation in (9.14), multiplied by $A\frac{dp}{dA}$, and the time derivative of the latter, multiplied by ρ. The result is a wave equation identical to (9.15) with velocity U in the place of pressure p, whose solution is a rigid propagation described by an arbitrary function $U(x \pm ct)$ dictated by the initial condition.

For reference, using the second equality (9.16) it is possible to estimate the celerity in aorta, where $E \approx 10^5$ N/m^2 and $s/D \approx 0.1$, between 3 to 5 m/s. Smaller vessels are relatively more rigid and celerity increases to above 10 m/s in peripheral arteries. The celerity formula (9.16) is often used to estimate the Young modulus of arteries, their stiffness, which is a pathological degeneration typical of aging. Celerity is measured by recording the pressure peak at different positions along the vasculature and it is obtained by the distance between the measurement points divided by the time difference between the passage of the wave.

The linear analysis of pressure pulse propagation presented here is based on several assumptions. A general nonlinear treatment is complicated and it is out of the present scope, more extensive mathematical analysis can be found elsewhere (Pedley 1980). However, it is instructive to mention how the approximations would affect the general solution (9.17).

First, the assumption of a fluid velocity much smaller than celerity, which led to neglecting the nonlinear term in (9.11), does not consider that the effective propagation velocity is $U \pm c$ and not just c. This means that the propagation velocity is not constant along the vessel, and the pressure wave does not propagate rigidly. Considering that in the circulation pressure and the velocity waves can be assumed as roughly in phase, the peaks of the pressure wave move faster than its trough, which gives rise to a sharpening of the front side of the pressure wave. Secondly, the celerity (9.16) is not a constant number, it changes with the vessel size and it is inversely proportional to the square root of the diameter D in the approximation given by the second equality in (9.16). Thus, the propagation is faster where the vessel is smaller and vice versa producing a further effect whereby the wave changes its shape during propagation. Ultimately, a non-zero friction would produce the attenuation of the wave that is smoothed out while propagating downstream.

9.4   Impulse Propagation at a Bifurcation

When a pressure wave reaches a bifurcation, it is partly transmitted downstream continuing the propagation into the daughter vessels and it is partly reflected backwards (see Fig. 9.4). Thus, at a bifurcation, there is a superposition between the incident wave (i) moving downward, the reflected wave (r) propagating upward and the two transmitted waves (t_1 and t_2) downward.

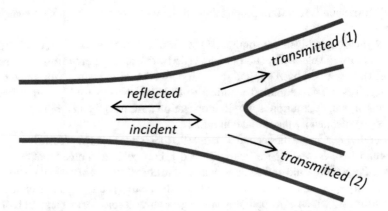

Fig. 9.4 Wave propagation at a bifurcation

In order to quantify this phenomenon, we can express the continuity of pressure at the junction, stating that pressure takes the same values when seen from the position of the different vessels

$$p_i + p_r = p_{t1}, \ p_i + p_r = p_{t2}; \tag{9.18}$$

and the conservation of mass that gives a relationship among discharges

$$Q_i - Q_r = Q_{t1} + Q_{t2}. \tag{9.19}$$

Assuming that the incident pressure p_i is known, there are 3 unknown pressure values and two Eqs. (9.18) that put them in relation. In order to move forward, we can consider the additional Eq. (9.19), and verify if it can be restated in terms of pressure. To this aim, we need to find a relationship between pressure and discharge. Consider a generic sinusoidal pressure wave $p(x, t) = F_p e^{i\omega(x-ct)}$ and the corresponding velocity wave $U(x, t) = F_U e^{i\omega(x-ct)}$; both are in the form (9.17) and represent a solution of the wave Eq. (9.15). Substitution of these expressions into any of the two Eqs. (9.14) gives the relationship $F_U = \frac{F_p}{\rho c}$. This indicates that, at least in a simple wave, the discharge can be related to pressure through

$$Q = \frac{A}{\rho c} p. \tag{9.20}$$

This relationship between flow and pressure is commonly expressed as

$$Q = \frac{p}{Z}, \ Z = \frac{\rho c}{A}, \tag{9.21}$$

introducing the concept of impedance, Z, that is a characteristic of a vessel.

Introduction of (9.21) in (9.19) gives, with (9.18), a system of 3 equations

$$\begin{cases} p_i + p_r = p_{t1}, \\ p_i + p_r = p_{t2}, \\ \frac{p_i - p_r}{Z_0} = \frac{p_{t1}}{Z_1} + \frac{p_{t2}}{Z_2}, \end{cases} \tag{9.22}$$

where the subscript 0 at the impedance stands for the parent vessel and the numbers 1 and 2 for the two daughters. Substitution of the first equations into the third gives a single equation relating incident and reflected waves

$$\frac{p_i - p_r}{Z_0} = \frac{p_i + p_r}{Z_1} + \frac{p_i + p_r}{Z_2}.$$

This can be rewritten by introducing the coefficient of reflection R

$$\frac{p_r}{p_i} = R, \ R = \frac{\frac{1}{Z_0}-\frac{1}{Z_1}-\frac{1}{Z_2}}{\frac{1}{Z_0}+\frac{1}{Z_1}+\frac{1}{Z_2}} = \frac{Z_1 Z_2 - Z_0 Z_2 - Z_0 Z_1}{Z_1 Z_2 + Z_0 Z_2 + Z_0 Z_1} ; \tag{9.23}$$

and a coefficient of transmission can also be obtained after substitution into the first equations in (9.22).

A perfect bifurcation would be able to transmit the pressure pulse downstream with no reflection, $R = 0$, which realizes when $\frac{A_0}{c_0} = \frac{A_1}{c_1} + \frac{A_2}{c_2}$. We know (remind Eq. (9.6)) that $A_1 + A_2 > A_0$, and that the celerity in a smaller vessel is usually higher, therefore in real bifurcation the reflection is effectively small, although not exactly zero. Thus, most of the pressure pulse is transmitted downstream and only in small part reflected upstream.

An important place where reflected waves can be effective is the pulse propagation along the aorta where it encounters the iliac bifurcation. This phenomenon is sketched in Fig. 9.5. As a starting model consider a simplified configuration where there are no side branches or geometric changes along the vessel. The incident pressure wave (i) starts from the aortic root and reaches the bifurcation after a time T (which takes typically a value about 0.1 s, given by the ratio between the length of the aorta, say something about 40 cm, and the celerity, say about 4 m/s). At this point, the reflected wave (r) travels backward from the bifurcation and reaches the aortic root after a time $2T$. Therefore, the pressure pulse that can be measured at the aortic

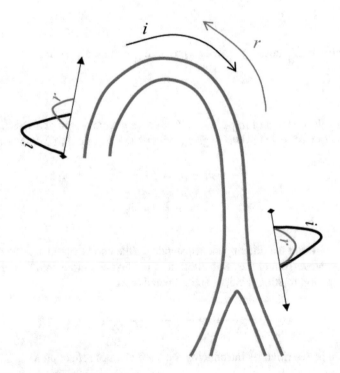

Fig. 9.5 Wave reflection in the Aorta

root results from the sum of the incident pressure wave, that is given by ventricular contraction and is made of a single impulse, plus the reflected wave that is similar to the incident wave, but lower in amplitude and delayed of $2T$. Physiologically, the backward traveling waves sustain the pressure at the aortic root after the initial impulse has passed and (with multiple reflections) ensures its slower decay during diastole. This sustained pressure helps to maintain the aortic valve closed and it is believed to help to provide allowance to the coronary flow during diastole. There are, however, other mechanisms involved and it is still unclear whether the role of reflection is fundamental or secondary. In any case, it is important to be aware that the time profile of pressure measured at any place does not reflect only the primary cause generating pressure (like ventricular contraction at the aortic root) as it also includes the contribution of reflected waves.

9.5 Collapsible Vessels

Arteries present a positive transmural pressure and typically operate under stretched conditions. There are, however, some other biological districts where internal pressure can become lower than the external value and the vessel be subjected to contraction. A contracted vessel maintains the circular geometry for small deformations only, then it undergoes a bending instability and collapses reducing sharply its area as shown in the generic "tube law" sketched in Fig. 9.6. A collapsed vessel gives high resistance to the flow and increased pressure losses.

Fig. 9.6 Tube law for collapsible vessels

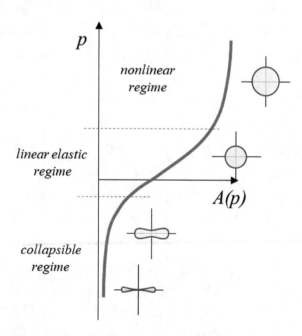

To exemplify possible implications of this fact, consider a vessel where flow starts with a given upstream transmural pressure p_0 at $x = 0$ that decreases downstream, for example following the Poiseuille law

$$p(x) = p_0 - \frac{128}{\pi} \mu \frac{Q}{D^4} x. \tag{9.24}$$

If the discharge Q increases, the pressure reduction is more pronounced along the vessels and the diameter decreases as well. Under certain conditions, when the upstream pressure is relatively low or the vessel is long enough, transmural pressure can become negative and, from the tube law, it can enter in a collapsible regime. A collapsed vessel further increases the pressure losses and enhances the level of collapse rapidly to reach a level that does not allow any increase of discharge.

This phenomenon is called "flow limitation" (Pedley 1980). It is common when the pressure difference is obtained by a reduction of the downstream value, like sucking from a straw capped at the top. It also occurs when the pressure is increased upstream to try pushing a higher discharge, because it may lead to more pressure losses than how pressure was increased upstream. Thus, the pressure gets further reduced downstream and leads to the collapse of the vessel that does not allow flow passage. Flow limitation typically occurs in airways, where one cannot blow more than a limited air rate otherwise the airways collapse, and blowing is reduced. It can also occur in male urination or in some long veins. This phenomenon was much studied in the giraffe jugular vein (Pedley et al. 1996), whose long neck represents of prototype for the analysis in a living system.

Part IV
Advanced Analysis of Separated Flow

Chapter 10
Vorticity and Boundary Layer Separation

Abstract Vorticity represents the underlying skeleton of fluid motion. The spatial distribution of vorticity characterizes the dynamic properties of fluid flow. When vorticity has a regular distribution, like when it is confined in the boundary layer next to the wall, the flow is regular and predictable. However, many flows are characterized by the instability of the boundary layer and formation of vortices; phenomena that give rise to complex motion and unpredictable trajectories. This chapter presents the basic elements of the dynamics of vorticity to provide a key for the understanding of fluid flow behavior in complex conditions. The flow pattern is related to the spatial distribution of vorticity, and the Navier–Stokes equation is expressed in terms of vorticity to identify the allowed evolution schemes. Based on these, the concept of boundary layer is restated in terms of vorticity, naturally leading to the process of vortex formation. Then the evolution of vortices is analyzed from their generation to their evolution up to dissipation. This chapter provides the foundation for a correct interpretation of blood flow when it develops complex behavior as can be found in cardiovascular sites of pathophysiological interest.

10.1 Vorticity and Irrotational Flow

The fluid velocity was assumed so far as the fundamental quantity for characterizing fluid motion. Velocity certainly is the most immediate and intuitive vector field to describe flows; however, it may not be able to evidence features of the underlying dynamical structure, like stresses, mixing or turbulence, that depend on velocity gradients. The weakness of a description based on velocity alone is particularly critical when the fluid motion features the presence of vortex structures. It will be clear shortly that *vorticity* is often a preferable fundamental quantity for the analysis of incompressible fluid dynamics.

Vorticity vector field was previously introduced through Eqs. (3.6) and (3.7); it is mathematically defined as the *curl* is of the velocity field

$$\boldsymbol{\omega}(\boldsymbol{x}, t) = \nabla \times \boldsymbol{v}(\boldsymbol{x}, t); \tag{10.1}$$

and represents the local rotation rate of fluid particles. More than that, it allows emphasizing the physical structure that hides behind the velocity field; it also provides a complete mathematical description of the flow and allows recovering the whole velocity field once the boundary conditions are imposed.

The interpretation of vorticity is particularly intuitive in two-dimensional flows when only the x and y components of the velocity field exist with no change along z. In this simple case, vorticity has only the z-component, perpendicular to the plane of motion, $\omega = \frac{\partial v_y}{\partial x} - \frac{\partial v_x}{\partial y}$, and physically corresponds to (twice) the local angular velocity of a fluid particle. In fact, while moving with the velocity of a particle, the presence of a positive vorticity corresponds to a vertical velocity increasing horizontally, $\frac{\partial v_y}{\partial x} > 0$, and a horizontal velocity decreasing vertically $\frac{\partial v_x}{\partial y} < 0$. It is easy to understand, see Fig. 10.1 (leftmost sketch), that this type of velocity variation represents a rotational motion on top of the particle translation. The general background behind this interpretation was previously described in Sect. 3.3.

The relevance of vorticity is not limited to local rotation. The spatial distribution of vorticity characterizes the different possible types of fluid motion. For this reason, vorticity is commonly considered the skeleton of the flow field and the fundamental quantity to describe the underlying flow structure. It is common to loosely describe a *vortex* as a region where fluid motion presents circular or swirling streamlines; however, it is more appropriate to define a vortex as a region of compact vorticity as shown in the central sketch of Fig. 10.1. The intensity of a vortex is normally measured by its *circulation*, normally indicated with Γ, that is the surface integral of the (normal component of) vorticity contained inside a certain region (the vortex

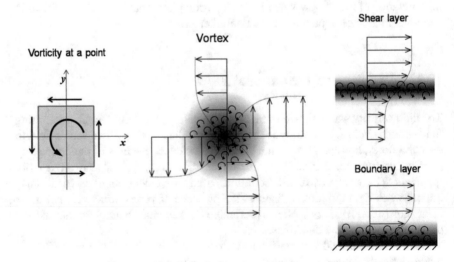

Fig. 10.1 Vorticity corresponds to the local rotation of a fluid particle. The spatial distribution of vorticity gives rise to different flow structures. An accumulation of vorticity in a compact region corresponds to a vortex; an elongated distribution of vorticity corresponds to a shear layer that, when it is adjacent to the wall, is a boundary layer

area) that, by the Stokes theorem (3.16), is equivalent to the circulation of velocity along a closed circuit surrounding the vortex area.

$$\Gamma = \int \omega \cdot n \, dA = \oint v \cdot ds. \tag{10.2}$$

The other fundamental flow pattern in fluid flow is the *shear layer*, an elongated layer of friction between streams moving with different velocities. Following the right sketches in Fig. 10.1, a shear layer is best described as a layer of vorticity, a *vortex layer*. Thus, the boundary layer discussed previously in Sect. 7.1 is a vortex layer adjacent to the wall that develops because of the velocity difference between the outer flow and the fluid attached to the wall for viscous adherence. The intensity of a vortex layer is measured by the difference of velocity, the *velocity jump*, commonly indicated by γ, between the flow above and below the layer; which is equivalent to the line integral of the vorticity along a line crossing the layer. Vortices and vortex layers are the fundamental vorticity structure in flow fields. Their different three-dimensional arrangements and their combinations give rise to the complexities of all evolving flows. The analysis of these vorticity elements allows a more intimate description of incompressible flows and provides guidelines for an intuitive understanding of their evolution (Kheradvar and Pedrizzetti 2012; Panton 2013).

The fundamental role of vorticity can also be appreciated from a mathematical perspective. It originates from the mathematical decomposition due to Helmholtz telling that any (sufficiently regular) three-dimensional velocity field can be expressed as the sum of two contributions: one *irrotational* component, that can be expressed as a gradient of a scalar potential ϕ, plus one rotational component expressed as the curl of a vector *stream function* field ψ, that has zero divergence $\nabla \cdot \psi = 0$. In general, one can write

$$v = \nabla \phi + \nabla \times \psi. \tag{10.3}$$

If one takes the curl of the velocity (10.3), the curl of a gradient is identically zero, and vorticity is contained in the rotational component only. The curl of (10.3) provides

$$\omega = \nabla \times (\nabla \times \psi) = \nabla(\nabla \cdot \psi) - \nabla^2 \psi; \tag{10.4}$$

where the second equality is obtained by standard identities between derivatives. The condition that the stream function has zero divergence gives the Poisson equation

$$\nabla^2 \psi = -\omega, \tag{10.5}$$

relating vorticity and stream function. An important point shown by this relationship is that the scalar potential ϕ does not contribute to vorticity as it represents the irrotational component of the velocity and, as such, it does not include rotations.

The Poisson Eq. (10.5) is a linear elliptic equation that can be solved by numerous means; the important concept to be learned is that the rotational component of the velocity field, $\nabla \times \boldsymbol{\psi}$, can be obtained from the knowledge of vorticity field.

In this respect, the irrotational component of the velocity field is particularly simple in incompressible flows. It participates in the conservation of mass only and it does not involve the equation of motion. This is immediate to verify by taking the divergence of (10.3). Recalling that the divergence of a curl is identically zero, the rotational field automatically satisfies the continuity Eq. (4.10) that becomes one equation for the potential only

$$\nabla^2 \phi = 0. \tag{10.6}$$

Equation (10.6) is a linear equation of the elliptic type, known as the Laplace equation. It is a homogeneous equation that has a unique solution driven by the boundary conditions only and can be solved by innumerable methods. It is important to remark that such velocity component represents an irrotational flow dictated by the continuity equation only, thus it can satisfy the instantaneous balance of mass but it does not include any balance of momentum because the Navier–Stokes equation and its evolutionary mechanism are not employed. In other words, the irrotational component in (10.3) can be obtained by kinematic congruence due to mass conservation whereas it does not include dynamic phenomena.

The implications for this analysis are important to describe the flow fields. A flow without vorticity is made of an irrotational velocity field only that can be specified without involving the balance of momentum. The equation of motion can be then employed, when required, to derive the pressure distribution corresponding to the known velocity field. In the case of irrotational flow, this can be performed with the simple Bernoulli equation for an ideal flow because energy dissipation is absent in an irrotational flow. In fact, the viscous term of the Navier–Stokes equation, $\nabla^2 \boldsymbol{v}$, which can be written for an incompressible flow as $\nabla \times \boldsymbol{\omega}$, is identically zero for a flow without vorticity.

The velocity decomposition is the key tool to recognize the role of vortices in a flow because only the dynamics of vorticity involve the balance of momentum. A vortex, as said, is a region where vorticity has accumulated; now we can also state that *a vortex is not necessarily a region exhibiting circulatory motion*. As shown in Fig. 10.2 (left picture), the velocity field corresponding to an isolated vortex is purely rotational, its streamlines rotate about the vortex and describe a circular motion. However, when an irrotational contribution adds on top of the same vortex, the irrotational velocity may modify the apparent vortex signature in terms of streamlines and the circulatory pattern remain hidden behind an irrotational contribution. To explain this point, let us consider the same vortex of Fig. 10.2 (left) with an additional uniform flow, a rigid translational motion from top to bottom that is evidently an irrotational component and does not affect the value of vorticity and of shear rate anywhere. The resulting flow fields are shown in Fig. 10.2 (central and rightmost panels) for increasing values of the uniform motion. The three fields of Fig. 10.2 present exactly the same vortex, the same gradients of velocity at all points. The rotational velocity field is always the

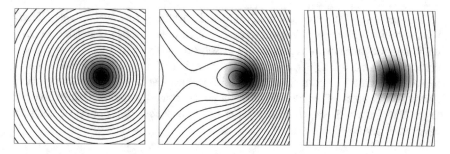

Fig. 10.2 A vortex is a region where vorticity has accumulated; it is not necessarily a region exhibiting circulatory motion. A flow made of a vortex only is made of circular streamlines (left panel). The streamlines are modified when a uniform vertical flow of moderate (center) ad high intensity (right panel) is added. In the three panels the vortex is unchanged, and so is shear in the flow

same, corresponding to the leftmost picture, only an irrotational flow is added to the others; nevertheless, from a superficial qualitative view in terms of streamlines, the underlying vortex may not be equally recognizable.

Fluid dynamics phenomena related to evolutionary dynamics, friction, dissipation, forces, boundary layer, vortex formation, etc., are dominated by the rotational part of the velocity field, while the irrotational velocity contributes in terms of transport and mass conservation only. Therefore, conceptually, a flow field can be described from the dynamics of the vorticity, plus an irrotational contribution to adjusting mass conservation accordingly to boundary conditions. This is why, when the flow field is not simple or mostly unidirectional, vorticity, and vortices in which vorticity organizes, are the fundamental quantity allowing a proper interpretation of flow evolution.

The vorticity is thus the fundamental quantity for describing a fluid flow. From the knowledge of the vorticity field only, the entire flow field inside a given geometry can be reconstructed. Technically, by inversion of Eq. (10.5), plus an irrotational component that is the solution of (10.6), both are linear elliptic equations that can be solved by numerous means (either analytic or numerical). It is, therefore, tempting to analyze the dynamics of fluid motion following the evolutionary dynamics of the vorticity itself; then solving the two linear equations for finding the complete velocity vector field that automatically satisfies mass conservation and obeys the boundary conditions. This is often useful because vorticity occupies only a small fraction of the flow field, and takes standard shapes that allow their immediate characterization.

The vorticity field has the further simplifying property that it obeys the same zero-divergence constraint of the velocity in an incompressible fluid: vorticity is a field with zero divergence (simply because the divergence of a curl is zero by definition)

$$\nabla \cdot \boldsymbol{\omega} = 0. \tag{10.7}$$

This means that the vorticity field cannot take arbitrary geometric shapes. Therefore, vorticity typically develops in terms of vortex tubes (whose associated velocity circulates the tube) or of vortex layers (associated with a difference of velocity, a shear rate, across the layer). Moreover, the total vorticity contained inside a vortex tube is conserved like the discharge in a tube of flow: a vortex tube cannot terminate abruptly, and must either be a closed ring or terminate by spreading into a vortex layer. Thus, we will use this knowledge to get a deeper description of complex cardiovascular flows.

10.2 Vorticity Equation

Vorticity is a vector field that follows a deterministic evolutionary law. Its mathematical expression can be immediately derived from the conservation of momentum: namely, the Navier–Stokes Eq. (5.34) that can be rewritten in terms of vorticity. Start by taking the curl of the Navier–Stokes Eq. (5.34)

$$
\nabla \times \frac{\partial \boldsymbol{v}}{\partial t} + \nabla \times (\boldsymbol{v} \cdot \nabla \boldsymbol{v}) = -\frac{1}{\rho} \nabla \times \nabla p + \nu \nabla \times \nabla^2 \boldsymbol{v}.
$$

Recall that (under sufficient regularity conditions) derivatives are linear operators and can be exchanged with the curl derivative operator, we get

$$
\frac{\partial \boldsymbol{\omega}}{\partial t} + \nabla \times (\boldsymbol{v} \cdot \nabla \boldsymbol{v}) = \nu \nabla^2 \boldsymbol{\omega}; \tag{10.8}
$$

where the pressure term, like any other conservative force, disappears because the curl of a gradient is identically zero. The second term in (10.8), which is nonlinear, requires some care. Consider the x-component of this term in a system of Cartesian coordinates

$$
\begin{aligned}
\nabla \times (\boldsymbol{v} \cdot \nabla \boldsymbol{v})|_x &= \frac{\partial}{\partial y}\left(v_x \frac{\partial v_z}{\partial x} + v_y \frac{\partial v_z}{\partial y} + v_z \frac{\partial v_z}{\partial z}\right) - \frac{\partial}{\partial z}\left(v_x \frac{\partial v_y}{\partial x} + v_y \frac{\partial v_y}{\partial y} + v_z \frac{\partial v_y}{\partial z}\right) \\
&= v_x \frac{\partial}{\partial x}\left(\frac{\partial v_z}{\partial y} - \frac{\partial v_y}{\partial z}\right) + v_y \frac{\partial}{\partial y}\left(\frac{\partial v_z}{\partial y} - \frac{\partial v_y}{\partial z}\right) + v_z \frac{\partial}{\partial z}\left(\frac{\partial v_z}{\partial y} - \frac{\partial v_y}{\partial z}\right) \\
&\quad + \frac{\partial v_x}{\partial y}\frac{\partial v_z}{\partial x} + \frac{\partial v_y}{\partial y}\frac{\partial v_z}{\partial y} + \frac{\partial v_z}{\partial y}\frac{\partial v_z}{\partial z} - \frac{\partial v_x}{\partial z}\frac{\partial v_y}{\partial x} - \frac{\partial v_y}{\partial z}\frac{\partial v_y}{\partial y} - \frac{\partial v_z}{\partial z}\frac{\partial v_y}{\partial z}.
\end{aligned}
$$

Recognize now that the terms in bracket are the x vorticity component, and use the continuity equation to group the second and third terms and fifth and sixth terms in the last line as follows:

$$
\nabla \times (\boldsymbol{v} \cdot \nabla \boldsymbol{v})|_x = \boldsymbol{v} \cdot \nabla \omega_x + \frac{\partial v_x}{\partial y}\frac{\partial v_z}{\partial x} + \left(\frac{\partial v_y}{\partial y} + \frac{\partial v_z}{\partial z}\right)\frac{\partial v_z}{\partial y} - \frac{\partial v_x}{\partial z}\frac{\partial v_y}{\partial x}
$$

$$-\frac{\partial v_y}{\partial z}\left(\frac{\partial v_y}{\partial y}+\frac{\partial v_z}{\partial z}\right)$$

$$= \mathbf{v}\cdot\nabla\omega_x + \frac{\partial v_x}{\partial y}\frac{\partial v_z}{\partial x} - \frac{\partial v_x}{\partial x}\frac{\partial v_z}{\partial y} - \frac{\partial v_x}{\partial z}\frac{\partial v_y}{\partial x} + \frac{\partial v_y}{\partial z}\frac{\partial v_x}{\partial x}$$

$$= \mathbf{v}\cdot\nabla\omega_x + \frac{\partial v_x}{\partial y}\frac{\partial v_z}{\partial x} - \frac{\partial v_x}{\partial x}\frac{\partial v_z}{\partial y} - \frac{\partial v_x}{\partial z}\frac{\partial v_y}{\partial x}$$

$$+ \frac{\partial v_y}{\partial z}\frac{\partial v_x}{\partial x} + \left(\frac{\partial v_x}{\partial y}\frac{\partial v_x}{\partial z} - \frac{\partial v_x}{\partial y}\frac{\partial v_x}{\partial z}\right);$$

where in the last passage we added a term, in brackets, that is equal to zero. Now group the terms properly as

$$\nabla\times(\mathbf{v}\cdot\nabla\mathbf{v})|_x = \mathbf{v}\cdot\nabla\omega_x - \frac{\partial v_x}{\partial x}\left(\frac{\partial v_z}{\partial y}-\frac{\partial v_y}{\partial z}\right) - \frac{\partial v_x}{\partial y}\left(\frac{\partial v_x}{\partial z}-\frac{\partial v_z}{\partial x}\right)$$

$$- \frac{\partial v_x}{\partial z}\left(\frac{\partial v_y}{\partial x}-\frac{\partial v_x}{\partial y}\right)$$

$$= \mathbf{v}\cdot\nabla\omega_x - \boldsymbol{\omega}\cdot\nabla v_x.$$

Reinserting this result into (10.8) gives the *vorticity equation*

$$\frac{\partial\boldsymbol{\omega}}{\partial t} + \mathbf{v}\cdot\nabla\boldsymbol{\omega} = \boldsymbol{\omega}\cdot\nabla\mathbf{v} + \nu\nabla^2\boldsymbol{\omega}; \tag{10.9}$$

which represents the Navier–Stokes equation expressed in terms of vorticity.

Despite the apparent mathematical complexity, the qualitative inspection of the terms in this equation permits to extract some important concepts regarding vortex dynamics. For example, it can be immediately recognized that the vorticity equation does not contain the pressure (or any conservative force like gravity). In fact, the distribution of pressure has no direct influence on vortex dynamics; on the contrary, however, pressure depends on vorticity that rules friction and energy losses and, if required, it can be obtained in the aftermath once velocity is known.

A first property that is read from the vorticity evolution equation is that if vorticity is zero at one instant it remains zero afterward. This important fact is immediately seen by inspection of Eq. (10.9), because all terms containing the vorticity spatial derivatives are identically zero when vorticity is zero thus the time derivative is also null and vorticity cannot depart from a zero value. This observation has the important implication that vorticity cannot be created inside the fluid; therefore, it can only be generated at the interface between the fluid and the boundary. This apparently simple fact is a fundamental element for the study of vortex dynamics: in incompressible flows vorticity does not appear spontaneously within the fluid, *the only place where vorticity can be created is at the boundary between fluid and tissue.* Indeed, the issue of the generation of vorticity, and vortex formation, in particular, is a key one and it will be extensively discussed in the next section.

Equation (10.9) tells that, once vorticity is somehow generated, it is subjected to few possible evolutionary phenomena. The primary one is that vorticity is transported with the flow as if it was a passive tracer (although not effectively passive, because velocity is related to vorticity itself). This phenomenon is provided by the two terms on the left-hand side of (10.9) that represent the Lagrangian time derivative of vorticity over fluid elements that move with the flow. The first term is the time variation of vorticity at the fixed position crossed by the particle; the second term gives an increase of vorticity when a particle points in a direction along which vorticity grows (i.e., when velocity is aligned with a positive gradient of vorticity). They take a form analogous to, for example, the first two terms in equation Navier–Stokes Eq. (5.34), describing the change of velocity (acceleration) on a moving particle. Therefore, summarizing, the vorticity of a fluid element moves with the local fluid velocity, like a tracer, but the value on such element can change its value in virtue of additional phenomena ruled by the two terms on the right-hand side of (10.9).

The first term represents the phenomenon of an increase of vorticity by *vortex stretching*. To visualize this, consider a small cylindrical element of fluid aligned with the vorticity vector, as sketched in Fig. 10.3. That cylinder of fluid translates

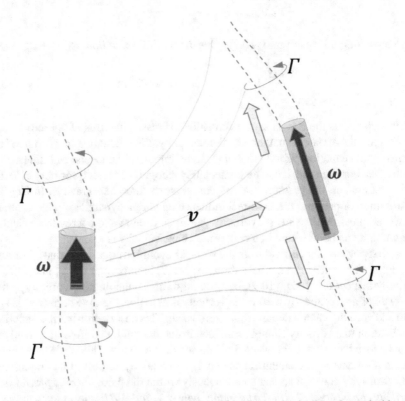

Fig. 10.3 A fluid element with vorticity is transported by the local velocity and stretched by the component of velocity gradient aligned with vorticity. This gives an increase of vorticity although circulation is conserved

with the local velocity as discussed above; moreover, the presence of a velocity difference between the two ends deforms the cylinder. Following Fig. 10.3, when the velocity component parallel to vorticity is lower at the base and higher at the top of the cylinder, as time proceeds the cylinder elongates, it is stretched by the velocity gradient (and shrunk in the transversal direction for the conservation of mass). Well, the vorticity vector behaves in the identical manner as material fluid, when fluid is stretched the vorticity vector is stretched as well and the vorticity value increases, whereas the cross-size reduces and the total amount of vorticity in the cross section, measured by the circulation Γ, is conserved. This phenomenon occurs in the presence of a velocity gradient in the direction of vorticity, i.e., $\boldsymbol{\omega} \cdot \nabla v > 0$. This term represents the stretching and turning of vortex lines as if they were lines of fluid. A further important aspect of this term is that it is exactly zero in a two-dimensional flow. In a two-dimensional flow, the vorticity is perpendicular to the plane of motion and there is no velocity gradient out of plane; therefore, vorticity stretching is intrinsically a three-dimensional effect.

Before turning the attention to the last term containing the viscous effects, let us recapitulate the dynamics of vorticity in absence of viscous effects. First, an element of fluid that contains no vorticity remains without vorticity afterward. This is the first of the three Helmholtz's laws for inviscid flow. Then, the vorticity is a vector that behaves like a small string element of fluid. It moves with the flow and it is stretched and tilted with it. This is essentially the second Helmholtz's law. The third law follows from the fact that vorticity is a field with zero divergence, and the total vorticity contained inside a vortex tube (or a vortex filament, when the tube is thin), measured by its circulation, is conserved along the filament while it moves with the flow.

The picture becomes extremely simple and intuitive in a two-dimensional flow, or in a motion that is locally approximately two-dimensional. In this case, the vorticity vector has a unique non-zero component perpendicular to the plane of motion, therefore it loses its vector character. Stretching is absent and vorticity and can be treated as a scalar quantity that is simply transported with the flow. The value of vorticity is stuck onto the individual fluid particles, vorticity simply accumulates into vortex patches, redistributes into the vortex layer, accordingly to the motion of fluid particles.

The last, viscous term in the vorticity Eq. (10.9) introduces the effects of friction and energy dissipation in terms of vorticity. The action of viscosity on vorticity is analogous to that of heat diffusion or diffusion of a tracer like ink or smoke. The spatial distribution of vorticity is smoothed out by viscosity; therefore, a sharp vortex reduces progressively its local strength while it widens its size in a way that the total vorticity is conserved. In general, the diffusion process is of simple interpretation. Like in any diffusive process, the rate of diffusion is higher in the presence of sharp vorticity gradients, therefore the magnitude of viscous dissipation becomes increasingly relevant where vorticity presents changes over short distances. This leads to the most important aspect of energy losses in fluid motion: *viscous dissipation is most effective at small scales*. Viscous diffusion, for example, gives rise to the annihilation of close patches of opposite sign vorticity. This has a peculiar consequence in three dimensions when two opposite-sign vortex filaments get in contact, the opposite-sign vorticity locally annihilates and oppositely pointing vortex lines (that cannot terminate into the flow) reconnect. The viscous reconnection phenomenon is the underlying mechanism leading to topological changes, metamorphoses of three-dimensional vortex structures, and increased dissipation by turbulence; for this reason, it will deserve further discussion later in this chapter when describing the interaction between vortices,

In summary, the dynamics of vorticity is made by its transport with the fluid elements, intensification by three-dimensional straining of such fluid elements, and smoothing by viscous diffusion. A dynamic that sees vorticity arranged into tubular and sheet-like structures ensuring a continuity of vortex lines. Some exemplary realizations of vorticity dynamics will be discussed later; before then, however, it is necessary to address the aspect of the generation of vorticity.

10.3 Boundary Layer Separation and Vortex Formation

As remarked above, in incompressible flows vorticity cannot be generated inside the fluid domain. Vorticity can only be created at the boundary in consequence of viscous adherence between the fluid and the surrounding tissue. Vorticity is produced because the no-slip condition applies at the interface between the fluid and the solid surface gives rise to a vortex layer adjacent to the boundary; such layer then progressively diffuses away from the wall through the viscous diffusion mechanism to produce a smooth boundary of vorticity at the boundary. The boundary layer thickness corresponds to the length at which the viscous diffusion penetrates from the wall into the flow, which is proportional to $\sqrt{\nu t}$ as seen in Eq. (7.12). The boundary layer was introduced in Sect. 7.1 as the region adjacent to the wall where the velocity rises from the zero value that it takes at the boundary to a finite value away from it. However, its interpretation as a vorticity layer is more intuitive for addressing vortex formation processes.

The boundary layer has a fundamental importance in fluid mechanics as it represents the *unique* source of vorticity in a flow field. In order to fix the ideas, consider a set of Cartesian coordinates at the wall, with x and y parallel to the wall and z perpendicular to it, at the wall the normal vorticity vanishes for the adherence condition and the others take a simple form

$$\omega_x = \frac{\partial v_z}{\partial y} - \frac{\partial v_y}{\partial z} = -\frac{\partial v_y}{\partial z},$$
$$\omega_y = \frac{\partial v_x}{\partial z} - \frac{\partial v_z}{\partial z} = +\frac{\partial v_x}{\partial z},$$
$$\omega_z = \frac{\partial v_y}{\partial x} - \frac{\partial v_x}{\partial y} = 0;$$

therefore, assuming x as the local direction of the velocity, the adherence generates the vorticity component ω_y perpendicular to it (as if the flow moves over wheels, with vorticity representing their axis) and vorticity generation begins, locally, as a two-dimensional phenomenon. It is also immediate to recognize that the components of vorticity at the wall correspond to the wall shear rate and, after multiplication with viscosity, to the wall shear stress (WSS, τ_0)

$$\begin{aligned} \tau_{0_x} &= +\mu\omega_y, \\ \tau_{0_y} &= -\mu\omega_x; \end{aligned} \tag{10.10}$$

which, we will show later, takes particular relevance in cardiovascular physiology and pathogenesis. Therefore, the wall vorticity is often employed interchangeably with wall shear rate (sometimes, given the constancy of viscosity, also with wall shear stress).

In small vessels, the thickness of the boundary layer is comparable to the diameter and fills the entire flow field. At such small scales, as found in arterioles and capillaries, viscous diffusion is the dominant phenomenon; vorticity is generated for adherence and quickly diffuses into the whole domain. In this case compact vortices, with rare exceptions, are absent. On the contrary, in large blood vessels or inside the cardiac chambers, the boundary layer often remains thin and it is capable to penetrate for diffusion over a small fraction of the vessel size. Indeed, until it remains attached to the wall, it has a relatively small influence on the flow and it only represents a viscous slipping cushion for the outside motion. However, under many circumstances, it happens that such a thin boundary layer detaches from the wall and enters into the bulk flow. This is the process of *boundary layer separation*, when thin layers of intense vorticity penetrate in the main flow and give a local accumulation of vorticity and eventually to the formation of compact vortex structures.

Boundary layer separation is key in the development of complex motion in large blood vessels. It occurs as a consequence of the local deceleration of the flow, which is normally associated to geometrical changes. The process of boundary layer separation is sketched graphically in Fig. 10.4. When flow decelerates, the upper edge of the boundary layer is subjected to deceleration as well and, because of incompressibility, when the longitudinal velocity decreases downstream the vertical velocity must increase moving away from the wall thus producing a local growth of the thickness of the vortex layer at the same location (Fig. 10.4 top picture). This tongue of vorticity is lifted and gets strained by the outside flow that has a higher velocity. Although vorticity is transported by the local velocity, it is also closely related to velocity and describes its spatial variations; therefore, as vorticity is strained away, the longitudinal velocity profile therein reduces and the vorticity at the wall below decreases as well. As this process progresses, opposite sign wall vorticity appears, and a secondary boundary layer develops below the separating shear layer (middle picture). The separation point at the wall, from where the separation streamline departs, corresponds to the place where vorticity is zero. The secondary vorticity is itself decelerated in its backward motion and is lifted up. Eventually, it cuts the connection between the original boundary layer and the separating vorticity that detaches and enters into the flow (bottom picture). During this process, it is important to remind that the local velocity transports vorticity but the latter is not a passive

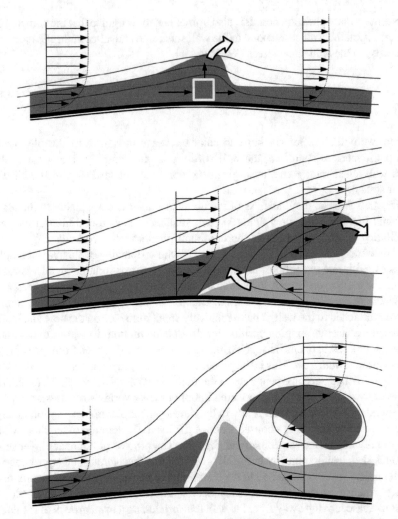

Fig. 10.4 Sketch of the boundary layer separation process. The dark gray indicates layers with clockwise vorticity, the light gray is counterclockwise; streamlines and velocity profiles are drawn. The deceleration of the flow produces a local thickening of the boundary layer due to mass conservation balance (upper panel). Such emerging vorticity is, therefore, lifted and transported downstream by the external flow (see arrows). A shear layer then extends away from the wall and produces a secondary boundary layer, with oppositely rotating vorticity (mid panel). The separated clockwise vorticity tends to roll-up while the secondary layer lifts up for the same initial mechanism, because it backward motion is decelerating (see arrows). Eventually, the separating vortex layer detaches from the boundary layer and becomes an independent vortex structure.The process of boundary layer separation involves a competition between the tendency of vorticity layers to become progressively sharper and their tendency to become thicker for viscous diffusion. Therefore, sharp, well-recognizable vortices are generated in large vessels where the role of viscous diffusion is smaller. Vice versa, in small vessels, the process of boundary separation is often inhibited by the dominance of diffusion that quickly smooths out the separating vorticity

tracer, it is made of velocity gradients that, when transported, alter the underlying structure of the flow itself affecting the rotational component of velocity in virtue of Eq. (10.5). Figure 10.4 shows qualitative velocity profiles and streamlines that develop in correspondence with the separating vorticity field.

Boundary layer separation is thus a consequence of the local deceleration of the flow. In other terms, separation develops in presence of an *adverse pressure gradient* (pressure growing downstream) that pushes from downstream and decelerates the stream. The most common way to have an adverse pressure gradient is that of a geometric change: a positive curvature of the wall, like an enlargement in a vessel. In this case, the velocity decreases, for mass conservation, kinetic energy decreases, and the value of pressure increases for the Bernoulli balance. Therefore, boundary layer separation develops when a vessel increases its cross-section as it may occur behind a stenosis, or at the entrance of an aneurism. An extreme case of geometric change is that of a sharp edge; this is often found at the entrance of a side-branching vessel, and certainly on the trailing edge of the leaflets of cardiac valves. In the case of sharp edges, the flow deceleration is so local that the position of boundary layer separation is definitely localizable at the edge. The vorticity that developed on the upstream side detaches at the sharp edge and leaves the tissue tangentially.

Geometric changes are not the unique possible source for the development of a flow deceleration. Immediately downstream of branch sucking fluid away from the main vessel, the velocity reduces and an adverse pressure gradient develops. Similarly, boundary layer separation develops for the so-called *splash* effect, when a jet reaches a wall and produces high-velocity streamlines that decelerate when they are deflected along the wall. Finally, the local flow deceleration can also be produced by previously separated vortices. A vortex that gets close to a wall gives rise to a localized increase (or reduction, depending on its circulation) of the flow velocity at the wall below, and a corresponding deceleration immediately downstream (or upstream). The vortex-induced boundary layer separation is a frequent phenomenon that may become particularly critical in some applications. In fact, the area of principal separation is often localizable and properly protected, whereas a separation induced downstream due to a previously separated vortex may occur at unexpected locations.

The separation of the boundary layer represents the starting phase of the vortex formation process. The featuring property of any shear layer, as shown previously with Fig. 10.1, is the difference of velocity between its two sides: the farther side of the shear layer that detaches from the wall moves with a speed that is higher than the side closer to the wall. Therefore, the separating shear layer curves on itself and eventually rolls up into a tight spiral shape. Now, during the rolling-up process, the distance between two successive turns of the vortex layer progressively reduces, with the closest neighboring turns at the center of the spiral. The viscous diffusion process smears out this tight spiraling structure into a compact inner core with a smooth distribution of vorticity. The roll-up and the formation of an isolated vortex behind a sharp edge obstacle are shown in Fig. 10.5. The degree of smoothness depends on the level of viscous diffusion relative to the strength of the flow that induces separation, namely it depends on the value of the Reynolds number.

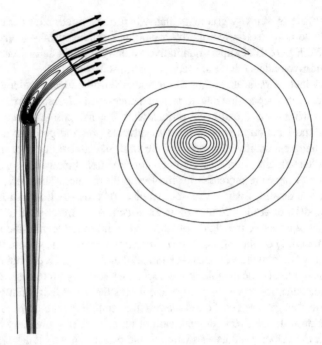

Fig. 10.5 Vortex formation from a sharp edge obstacle. The shear layer separates from the upstream "wetted" wall and rolls-up into a spiral. The tight turns in the inner part of the spiral spread for viscous diffusion into the inner core of the formed vortex

The vortex formation from a smooth surface is still described by the picture given above, where a few additional elements of complexity can be emphasized. First, the actual position of separation depends on the local flow structure; it cannot be preliminarily identified and may even change during time. Furthermore, given that the vortex formation process affects the surrounding flow field, it particularly influences the progress of separation from a smooth surface where the separation point can largely vary. Furthermore, separation from smooth surfaces is subjected to a more direct interaction between the forming vortex and the nearby wall where the viscous dissipation effects normally support the formation of smoother vortex structures.

One typical example of the external separation from the smooth surface of a bluff body is shown in Fig. 10.6 featuring the formation of oppositely rotating vortices from the two sides of a circular cylinder. In such an example, vortices interact and influence the opposite separation process eventually producing a sequence of alternating vortices, known as the von Karman vortex street, that is usually found behind bluff bodies. The development of alternating vortices is quite a common phenomenon when previously separated vortices may influence vortex formation in nearby regions. It is also present, with some differences, in internal flows when a vortex formed on one side of a vessel creates a vortex-induced separation on a facing wall. That, in

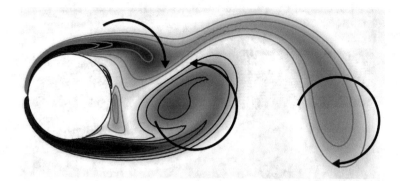

Fig. 10.6 Formation of vortices behind a circular cylinder. Oppositely rotating vortices separate from the two sides of the body in an alternating sequence. The previously separated clockwise vortex detached from the upper wall translated downstream, a counter-clockwise vortex has been formed from the lower wall, and a novel clockwise vortex is under formation from the wall above

turn, may induce a weaker further separation in a sort of wavy pattern extending and decaying downstream.

The internal separation, with the following formation of a vortex inside of a vessel is, in general, a smoother phenomenon because the presence of confining walls does not allow vortices to grow into large structures, keeps vortices more constrained to smaller sizes, and is more affected by viscous diffusion. Nevertheless, the presence of a vortex inside a vessel may give rise to alteration for the entire flow. It has a blocking effect that locally deviates the streamlines modifying the wall shear stress distribution, possibly producing further separations. It changes the unsteady pressure drop and in a branching duct, it may affect the division of the relative flow in the daughter vessels. An example is given in Fig. 10.7 that reports the vortex formation in the bulb of a carotid bifurcation. During the systolic acceleration, the boundary layer separates tangentially from the common carotid artery and develops a smooth roll-up within the bulb close to the nearby wall. During deceleration, the formed vortex locally affects the wall shear stress inside the bulb with multiple opposite sign wall vorticity. It has a blocking effect that deviates the streamlines at the entrance of the internal carotid artery into a faster jet. It produces secondary vortex-induced separation inside the internal carotid; it eventually (not shown in the picture) gives a secondary vortex formation and a further small separation little downstream.

A peculiar phenomenon associated with the vortex formation process can be outlined when the flow enters from a small vessel into a large chamber forming a jet whose head is the forming vortex. Here, after the very initial roll-up phase, a measure of the length of such a jet is given by the product of Ut where U is the velocity at the opening and t is the time. In this case, it is enlightening to define a dimensionless *vortex formation time*, *VFT*, as the ratio of the jet length with respect to the diameter of the opening D

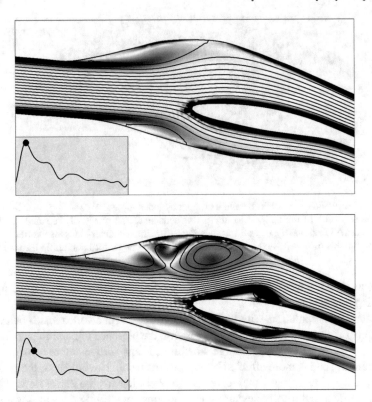

Fig. 10.7 Formation of vortices in a model of a carotid bifurcation. The accelerating systolic flow (upper panel, at peak systole) leads to a smooth boundary layer separation at the carotid bulb. After the peak (lower panel) the vortex just formed at the bulb either interacts with the bulb boundary layer creating multiple small vortices, and gives rise to a vortex-induced secondary separation in the oppositely facing wall of the internal carotid artery. The same phenomena in a much weaker version are noticeable also on the opposite side at the entrance in the external carotid artery

$$VFT = \frac{Ut}{D}. \tag{10.11}$$

The formation time represents a dimensionless number that characterizes the progression of vortex growth and allows a unitary description under different conditions. In reality, the definition of formation time has a more profound physical meaning. The separating shear layer has a strength given by the jump of velocity between its two sides, given approximately by U, and translates downstream with a velocity that is something like the average of velocity on the two sides of the layer, which is $U/2$, thus it feeds the circulation Γ of the forming vortex at a rate

$$\frac{d\Gamma}{dt} \cong \frac{1}{2}U^2. \tag{10.12}$$

The formation time thus also represents the dimensionless measure of the vortex strength, the circulation $\Gamma \cong \frac{1}{2}U^2 t$, normalized with UD.

The definition (10.11) can be extended to the case when either U or D vary during time, by integration of the ratio U/D during the period of vortex formation.

$$VFT = \int \frac{U(t)}{D(t)} dt.$$

10.4 Three-Dimensional Vortices

The vortex formation process described above is given in terms of two-dimensional pictures. It allows an immediate and intuitive understanding of the fundamental phenomenon because the initial phase of any vortex formation process is, with rare exceptions, locally two-dimensional and the three-dimensional organization of the vorticity enters into play at some later stages.

The simplest case of three-dimensional vortex formation is that from a circular orifice, in that case, vortex formation has a circular symmetry and the forming three-dimensional vortex tube has the shape of a ring. Vortex rings are well-known objects in fluid dynamics as they are easily generated using a piston-cylinder apparatus. A vortex ring is a stable vortex structure, it has an axial symmetry and vortices with a shape close to a ring also tend to the axisymmetric shape by an internal homogenization. Because of their stability, vortex rings are often encountered in nature, including when puffing smoke out of the mouth.

Figure 10.8 shows one instant during the formation of a vortex ring behind a circular orifice. The vorticity distribution on a transversal section (left panel) shows the shear layer separating from the orifice that eventually rolls-up into the jet head; however, this planar picture corresponds to a three-dimensional vortex structure that is more difficult to represent on paper. The vortex ring corresponding to the vortex core is shown in the same picture (right panel) to emphasize the main element of the three-dimensional vortex. In general, however, there is some ambiguity on the effective delineation of a vortex boundary. This is not a big issue in two-dimensional systems when the entire vorticity field can be shown in color scale on the picture plane and the different elements of the vortex structure are immediately recognized, from the separating shear layer, to the rolling-up spiral, to the vortex core. This is also not an issue in this simple case that presents an axial symmetry and vorticity has only the azimuthal component: this flow is conceptually planar. Indeed, its three-dimensional representation, on the right panel of Fig. 10.8, certainly contains less complete information, and the choice of the vortex core boundary severely influences the three-dimensional structure that is eventually visualized.

Despite their simplicity, vortex rings represent an instructive example as they contain the first seed of three-dimensional vortex dynamics. A vortex ring presents

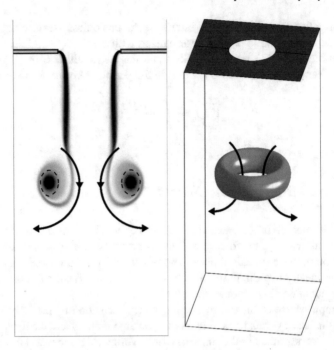

Fig. 10.8 Formation of a vortex ring from a circular sharp orifice. Left panel: distribution of vorticity on a transversal cross-cut; the vortex core is indicated with a dashed line. Right panel: three-dimensional view of the vortex ring core

a *self-induced velocity* that is due to the curvature of the vorticity lines (lines everywhere tangent to the vorticity field), an effect that does not exist in two dimensions. This follows from the relation between velocity and vorticity because a vortex line corresponds to a velocity field that rotates around the line; therefore, when vorticity is arranged in the form of a curved vortex tube, the associated rotation also induces a translation of the curved tube itself. Thus, once formed, a ring continues to translate downstream for its own self-induced velocity field. Such a self-induced velocity gives rise to a peculiar limiting process of three-dimensional vortex formation: during its formation, the vortex ring is continuously fed by the rolling-up shear layer separating from the orifice edge, therefore its circulation grows and the self-induced translation velocity of the vortex ring rises and exceeds that of the separating shear layer that is not more able to feed the vortex. At this point, the primary vortex detaches from the layer behind with a phenomenon known as *pinch-off*, at the same time the newly separated vorticity rolls-up in the wake of the main vortex, a phenomenon that is shown in Fig. 10.9. This limiting process occurs for a critical value of the vortex formation time that is about $VFT_{cr} \cong 4$. Above this limit, the vortex ring cannot grow as a unique structure and multiple vortices develop in its wake developing higher dissipation. It was found that the VFT in the human heart is close to this optimal limit and it decreases in diseased hearts (Gharib et al. 2006).

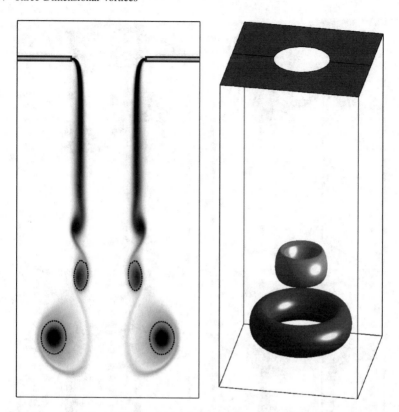

Fig. 10.9 Formation of a multiple vortex ring from a circular sharp orifice with a vortex formation time higher that 4 presenting the pich-off. Left panel: distribution of vorticity on a transversal cross-cut; the two main vortex cores are indicated with a dashed line. Right panel: three-dimensional view of the vortex rings core

The case of vortex ring formation after a circular opening represents the simplest case of three-dimensional vortex formation. Let us move forward and consider the flow across an orifice with a slender shape. In this case, the opening has a variable curvature and the separating vortex ring will not be circular and rather present a variable curvature. Therefore, the self-induced velocity, which is proportional to the curvature, will be different along the vortex tube and will progressively further deform it. When this deformation becomes high enough, the compact tubular vortex structure becomes unstable and breaks down into smaller elements, that in turn deform into even smaller ones, until they become small enough to be dominated by dissipation for viscous effect. One exemplary case of the three-dimensional vortex formation from a non-circular opening is shown in Fig. 10.10 where the vortex ring develops from an orifice made by two half circles connected by straight edges (Domenichini 2011). The behavior described above is rather common for vortex formation from a three-dimensional geometry; boundary layer separation gives rise to irregularly

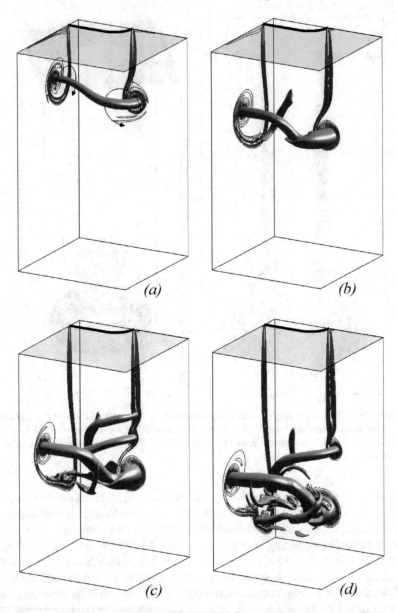

Fig. 10.10 Three-dimensional vortex formation from a slender orifice at four instants in sequence. One quarter of the entire space is shown for graphic clarity (allowed by symmetry); the vorticity contours are reported on the side planes to help understating the three-dimensional arrangement of the principal vortex filaments. In the initial phase, the formed vortex loop presents a variable curvature and deforms because of the different self-induced translation speed; this leads to further deformations until the vortex structure loses its individuality and becomes a set of entangled three-dimensional elements that rapidly dissipate for viscosity (credit: Domenichini. J Fluid Mech 2011;666:506, with permission by Cambridge University Press)

shaped vortex structures, which become unstable and break down into progressively smaller structures that eventually undergo to rapid energy dissipation.

The three-dimensional vortex formation from smooth surfaces, after a constriction like a stenosis or in a vessel enlargement, introduces additional elements of complexity that do not allow drawing a simple unitary picture of the involved phenomena. The initial instants following boundary layer separation and initial roll-up are essentially two-dimensional with a moderate influence from the three-dimensional structure. Afterward, the three-dimensional development leads to widely different results depending on the separating geometry, the interaction with the nearby walls, and with other surrounding vortices.

Before moving further ahead in this topic, it is important to remark that the vortex formation process is not just a kinematic adjustment of the flow but it has dynamic consequences. The generation of a vortex is associated with the development of a force on the walls from where the vortex originates; this "vortex force" is given by the rate of growth of the vortex impulse (Saffman, 1992)

$$F = \frac{d\boldsymbol{I}}{dt}, \quad \boldsymbol{I} = \rho \int \boldsymbol{\omega} \times \boldsymbol{x} \, dV; \tag{10.13}$$

remarking that the integral is non-zero only where the vortex is formed, thus vorticity changes during time. To clarify this point, consider the case of generating a vortex ring, which represents the roughly early stage of many three-dimensional vortex formation processes. The vortex impulse of a vortex ring of circulation Γ and radius R has only the component directed along the vessel axis, say x, (perpendicular to the plane containing the ring) which is $I_x = \rho \Gamma \pi R^2$; therefore, given that the radius does not vary or varies very slowly during the formation process, the force is proportional to the rate of growth of the vortex circulation

$$F_x \cong \rho \pi R^2 \frac{d\Gamma}{dt} \cong \frac{\pi}{2} \rho R^2 U^2, \tag{10.14}$$

which, using (10.12), turns out to be proportional to the square velocity across the orifice. This vortex force is due to the unsteadiness of the formation process. In a pulsatile flow, the vortex force (10.13), or (10.14), produces a continuous hammering localized onto the region of the tissues where the vortex develops.

10.5   Vortex Interactions with Other Vortices and with Walls

The ideal vortex formation picture described above is complicated when two or more vortices come nearby each other, because they likely interact in an intense and irreversible manner. The interaction of vortices involves many different and very complicated phenomena. In the simple case of two-dimensional vortices that come in a close encounter, they reciprocally induce a rotation velocity for each other. When such vortices have the same sign, they rotate together one around the other, winding up one over the other to eventually merge into a single larger one made by the sum of them. On the contrary, two vortices with opposite circulation, a vortex pair, translate together for the self-induced velocity (similarly to what a vortex ring does) along a straight or curved path depending on their relative strengths. Again, the differential velocity inside every single vortex produces the winding up of one's vorticity on other, however, such vorticity strips are of opposite sign and do not merge rather they annihilate each other and reduce the individual vortices' strength. When opposite vortices are very close they give rise to a single vortex with circulation equal to the algebraic sum of the original circulation, thus they annihilate when the circulations are equal and opposite.

The interaction between three-dimensional vortex structures occurs prevalently between two oppositely rotating portions of vortex tubes (because they are more likely driven one toward the other, while concordant 3D tubes tend to separate) and begins with the local interaction between the closest portions. One example of the interaction between two identical vortex rings is shown in Fig. 10.11. Initially, the local interaction is approximately two-dimensional: the nearby oppositely rotating tubular elements induce the velocity of each other and try to translate away. This produces a local stretching of the three-dimensional vortex tube, a stretching that accelerates while the tubes become closer. In addition, the close vorticity elements also tend to locally wind-up with one another. The interacting structures develop increasingly small scales until viscous diffusion becomes a dominant effect, at this point the *reconnection of vortex lines* occurs: adjacent opposite vorticity is annihilated by dissipation and the vortex tubes tend to fuse one onto the other.

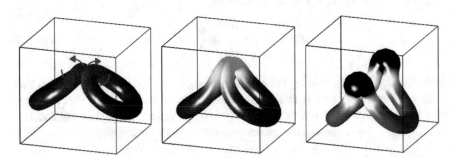

Fig. 10.11 Vortex interaction between two identical impacting vortex rings; the brightness of the filament indicates the strength of the corresponding vorticity. When oppositely rotating vortex tubes get close, they produce a local vortex stretching due to the self-induced velocity (from left to central panels). During stretching, the boundary between the vortices becomes locally sharper until the filaments fuse one into the other for viscous effect (from central to right panels). After vortex reconnection a new structure is formed, typically its geometry is irregular, the vortex is often unstable and short lived

The interaction between two identical vortices, like that shown in Fig. 10.11, may result in a complete vortex reconnection and a new vortex tube. More often, however, one vortex is stronger than the other is, only part of their tubular structures reconnects and such incomplete reconnection gives rise to new structures with a complex branched topology of vortex lines (see also Fig. 10.10 where some vortex reconnection occurs). In general, the vortex structures resulting from the fusion of interacting vortices typically present a very irregular geometry. Differential curvatures, which give sharply variable self-induced velocity and local motion, and differential vorticity strength, which give axial flow along the tube. These are all elements that tend to rapidly further deform the vortex, produce further reconnections and give rise to smaller vorticity structures. In other terms, an irregular three-dimensional vortex structure is overall unstable, tends to destroy itself, and it is short lived. The more a vortex is regular, like a vortex ring, the more it remains coherent and lasts longer.

Vortices also interact with the nearby walls; a phenomenon that is particularly relevant in closed geometries like the cardiovascular chambers. The vortex-wall interaction can be subdivided into two different phenomena: the *irrotational interaction*, which is a consequence of the wall impermeability; and the *viscous interaction* with the vorticity that develops at the wall for viscous adherence. Let us consider the two effects separately.

The first, irrotational interaction is due to wall impermeability only, and occurs before considering any generation of vorticity at the wall for adherence. In that scheme, when an isolated vortex, that would induce a rotary motion with circular streamlines, approaches an impermeable wall, the streamlines must deform to avoid crossing the boundary. As new vorticity cannot appear, the modification of the flow field due to the presence of the wall can only be imputable to the development of an additional irrotational field, expressed as $v_{irr} = \nabla\phi$, which satisfies the condition of impermeability at the wall. This additional flow is given by solution Laplace Eq. (10.6) with condition that the additional velocity cancels the original velocity, say v_0, perpendicular to the wall, $\frac{\partial\phi}{\partial n} = -v_0 \cdot n$. The solution of this linear equation is unique. The same solution can also be constructed considering symmetry arguments allowing a more immediate perception of the modification induced by the presence of the wall. Consider the flow that would be induced by an *image vortex* placed below the wall. We remind that the velocity induced by a vortex in a region away from it is also irrotational. With reference to Fig. 10.12 (left panel), consider then the irrotational flow that would be induced by an image vortex identical and of the opposite sign of the real one, placed symmetrically below the wall. Such an image vortex gives an irrotational flow whose velocity perpendicular to the wall that is opposite to that of the real vortex, and thus ensures that the fluid does not penetrate into it, it, therefore, corresponds to the sough correction induced by the presence of the wall. This representation allows a more immediate understanding of the effect induced by the correction. The velocity parallel to the wall has the same sign of that due to the real vortex and therefore the velocity adjacent to the wall increases (*splash effect*). In addition, the image vortex also induces a velocity to the real vortex that

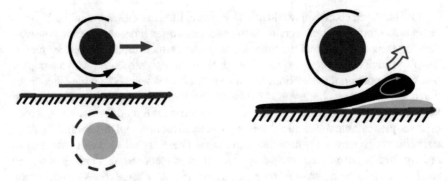

Fig. 10.12 The interaction of a vortex with the wall produces two separate effects. First (left panel), the condition of impermeability is satisfied by a distortion to the vortex-induced flow that is equivalent to having an opposite vortex placed symmetrically below the wall. The presence of such a "image" vortex increases the tangential velocity next to the wall, and induces a translation velocity to the otherwise still vortex. The second effect (right panel) is due to viscous adherence, the development of a boundary layer and eventually a vortex-induced separation

accelerates or decelerates (depending on the direction of the circulation) with respect to the background flow because of this *image effect*. For example, a (clockwise) vortex that just formed from a wall underneath is decelerated by the image below the same wall, while it accelerates when it approaches a wall on the opposite side.

The second phenomenon, additional to the image effect, that arises when a vortex gets near to a wall corresponds to the development of a vortex-induced boundary layer because of the viscous adherence. A vortex plus its image creates a local velocity gradient along the wall, acceleration followed by deceleration. This perturbation, as previously discussed, may give rise to a vortex-induced boundary layer separation and to the formation of secondary vortices as sketched in Fig. 10.12 (right panel). Such secondary vortices are of opposite sign and weaker than the main one; thus, they tend to wind up around the main vortex giving rise to close interaction of opposite sign vorticity that partly annihilate. The development of such thin shear layers during the interaction is counterbalanced by viscous spreading that is more intense when geometric scales get smaller, thus the intensity of the interaction depends on the strength of the vortex, its closeness to the wall, and size of the region, i.e., it depends on the value of an appropriately defined Reynolds number.

When the vortex-boundary interaction described above applies to a tract of a three-dimensional vortex tube, it applies first to the portion that is closer to the wall and eventually affects the following three-dimensional dynamics. First, the image effect gives a local change in the velocity that induces stretching and deformation of a vortex filament. Second, when the vortex gets closer, it eventually interacts directly with the vortex-induced vorticity distribution. This is an interaction between oppositely circulating vorticity. It gives rise to the local wind-up of the wall vorticity around the approaching vortex and to reconnection with its vortex lines. Eventually, the approaching vortex crops for the partial annihilation of the region closer to the wall, which unbalances the three-dimensional vortex structures that tend to rapidly further deform and develop small structures near the wall that are eventually dissipated. As a result, the interaction of a coherent vortex with a wall can give to change in the topology of the vortex or, more often, to a progressive dissipation of the entire structure.

10.6   A Further Account to Turbulence

Let us enter smoothly into the physics of turbulence by deepening a little further the concept introduced in Chap. 8. We've said here that vortices form for the separation of the boundary layer creating structures driven by the large-scale geometry of the fluid domain. After vortex formation, the vortex can undergo instabilities or interact with other vortices and nearby walls. These phenomena deform and break the overall, *large-scale* geometry of the vortex loops that, after sequences of instabilities and reconnections, eventually transform the original vorticity into several irregular small structures. Such *small-scale* elements present sharp velocity gradients, high viscous friction, and are rapidly dissipated.

Physically, on average, large scales vortices of the size of the containing geometry are continuously formed; these large vortices are unstable and produce progressively smaller flow structures until they are small enough to be dominated by viscosity and dissipate. From a high-level perspective, the resulting vortex-dominated flow witnesses the simultaneous presence of continuously generated large structures with others of all intermediate sizes developed by the instability of larger ones, and from these down to the smallest vortices dominated by viscosity. A measure of the complexity of such a flow field can be provided by from the amount of such contemporary vortices, measured by the ratio between the largest scale, say L, given by the size of the surrounding geometry (the diameter of the orifice, or the size of an obstacle, for example), and the smallest friction-dominated one, that we indicate with η. When L is comparable to η, the flow is a regular one because the generated vorticity is immediately smoothed out by viscosity; this is what happens in small vessels. On the other end, when L is much larger than η, the flow presents fluctuations over a large number of intermediate scales from L to progressively small sizes up to the smallest scale η. The order of magnitude of this complexity was be estimated for the case of a statistically steady turbulence on the basis of the phenomenological theory due to Kolmogorov in 1941 (Davidson 2004; Frisch 1995).

A flux of energy, indicated with ε, is injected in the flow at a scale close to L, say the energy forming a vortex behind an orifice. We have said that when $L \gg \eta$, the originally generated large vortex breaks down into smaller vortices, which in turn then break down into smaller ones and so on. In other terms, the energy rate responsible for the generation of the large vortex is transferred to smaller scales with essentially no dissipation until the size of the vortices is small enough and energy is eventually dissipated into heat by the smallest vortices that approach the viscous scale η. Therefore, in a statistical balance, the incoming energy rate ε corresponds to the rate of energy dissipation. It can be hypothesized that at small enough scales the statistical properties of turbulence become locally uniform and isotropic, independent of the details of how turbulence was generated. In this picture, the properties of the fluctuations of velocity depend only on the energy rate ε that arrives from large scales and is transferred to the small scales. This homogenization allows drawing a simple picture based on dimensional arguments, the rate of injection of kinetic energy (per unit mass) in the large scales is dimensionally given by a velocity square divided by a characteristic time. It can be estimated as proportional to the kinetic energy to U^2, being U a characteristic velocity, divided by the time, $\frac{L}{U}$, needed to cover the entire domain of size L. Thus, the energy rate can be estimated (as an order of magnitude) by

$$\varepsilon \sim \frac{U^3}{L}. \tag{10.15}$$

In such statistically steady, spatially uniform and isotropic condition, the viscous scale will depend solely on the amount of energy flux and by viscosity

$$\eta = f(\varepsilon, \nu). \tag{10.16}$$

This is a dimensional equation involving two units. By dimensional analysis, it is immediate to obtain the estimate, up to a multiplicative constant set to unity,

$$\eta \sim \left(\frac{\nu^3}{\varepsilon}\right)^{\frac{1}{4}}. \tag{10.17}$$

The viscous scale defined by (10.17) is also called the Kolmogorov scale. As said above, the degree of complexity of turbulent flows is represented by the amount of interleaving scales, which can be estimated by the ratio

$$\frac{L}{\eta} \sim \frac{L\varepsilon^{\frac{1}{4}}}{\nu^{\frac{3}{4}}} \sim \left(\frac{LU}{\nu}\right)^{\frac{3}{4}} = Re^{\frac{3}{4}}; \tag{10.18}$$

which is proportional to the Reynolds number at the power $\frac{3}{4}$. It again represents the measure of a separation between the scale of available energy and the viscous dissipation scale.

As a further remark, these estimates demonstrate how the Navier–Stokes equations, which do not allow general analytical treatments, may be difficult to be tackled even by numerical approaches when turbulence develops. Numerical solutions are based on a spatial discretization and it requires accuracy up to about the Kolmogorov scale to possibly reproduce the details of dissipative phenomena. Therefore, the space spanning the entire length of interest, whose size is proportional to L, must be sampled with a resolution about the size η. Thus, the number of sample points along any spatial direction must be about L/η and the total number of points required to sample a three-dimensional volume of fluid moving in a turbulent regime is something like

$$N_{3D} \approx \left(\frac{L}{\eta}\right)^3 = Re^{\frac{9}{4}}. \qquad (10.19)$$

The estimate (10.19) sets a limit to the actual feasibility of a comprehensive description of turbulent flows at a large Reynolds number. Due to this limitation, turbulence literature was mainly based on the solution of the Reynolds equations, or different versions of them based on different averaging/filtering (Sagaut 2006) introducing a "closure" model for the unknown terms appearing therein as discussed in Sect. 8.2. All such closure methods are, however, approximate and their reliability is limited to relatively simple flows. As a result, turbulence remains an open challenge and it is important to build a physical picture of possible turbulent phenomena in flows of interest.

In general, we may think of turbulence as a system of entangled and inter-acting vortex elements of disparate sizes. Ranging from the large size generated by the boundaries to the smaller size where the flow is smoothed out by viscous effects. Understanding that turbulence is generated by a sequence of interacting three-dimensional vortices allows its description in terms of the *energy cascade* described above. An external energy input (like a pressure difference across a valve) pushes a fluid across an orifice or along an irregular vessel bend. The flow thus generates energetic vortices whose size is comparable with that of the container. These large vortices interact and produce smaller eddies, which further interact producing turbulent eddies of a progressively smaller size capable to dissipate kinetic energy into heat. At the lower end of this energy cascade, very small eddies are entirely dissipated and do not generate anything smaller.

An increased friction between fluid elements and enhanced energy dissipation with respect to regular fluid motion characterizes turbulence. In fact, the development of turbulence is the strategy used by fluids to dissipate excess energy. When a fluid motion presents a large kinetic energy (high velocity), the fluid may be unable in a regular motion to maintain equilibrium between the external energy source and viscous dissipation, in that case, regular flow is unstable and instabilities increase the particle paths by developing swirling motions and small scales with higher shear rate to increases viscous dissipation up to equilibrium regime. The Reynolds number represents, through (10.18), the ratio between the kinetic energy introduced in the large scales, proportional to ρU^2, and their ability to dissipate with shear stress, grossly estimable as proportional to $\rho \nu U/L$. When the Reynolds number increases above a certain threshold, the regular flow becomes unstable and turbulence appears creating smaller scales to enhance dissipation. That's why every realization of flow motion presents a critical value of the Reynolds number above which the motion enters into turbulence.

In the cardiovascular system, turbulent flows are rarely encountered. The largest scales of motion achievable in the arterial network cannot exceed the vessel size, of a few centimeters at most. The Reynolds number is normally well below one thousand, with the exception of the largest vessels. The flow in the ascending aorta and, sometimes, in the cardiac chambers can reach the values of the Reynolds number up to some thousands. When turbulence develops, it is weak turbulence with an energetic level that does not influence dramatically the main dynamics. It should be remarked, however, that cardiovascular flows are not steady and some changes in the picture above are expected. In unsteady pulsatile flows, when the Reynolds number is high enough to suggest the presence of turbulence, the highest levels of turbulence are recorded during the deceleration after the peak of the flow. In fact, the flow received energy during the phase with high velocity and has to dissipate such energy during deceleration; the fast-moving fluid particles get closer approaching the preceding one, and therefore, the energy-filled fluid enhances instability phenomena during deceleration that supports turbulence. Vice versa, during acceleration fluid particles gain energy while moving away from each other and the flow is more stable.

The most frequent appearance of turbulence in the cardiovascular system occurs in the aortic artery that, on the other end, is the main artery and is involved in numerous pathologies. The flow across the tri-leaflet geometry of the aortic valve provokes a rather complex three-dimensional vortex formation that, associated with the large Reynolds number (roughly from 3,000 to beyond 10,000 at peak systole), produces weak turbulence. Weak turbulence may also develop during the diastolic filling of the left and right ventricles when the jet across the mitral or tricuspid valve, respectively, gives rise to the formation of three-dimensional vortex structures that interact with the surrounding ventricular tissues.

Boundary layer separation, vortex dynamics, and weak turbulence represent key elements in the interaction between fluid flow and surrounding tissues in large vessels. Understanding these fundamental phenomena is necessary to allow proper interpretations of fluid dynamics in cardiovascular regions of interest. They are particularly relevant for pathological developments and will be discussed in the next chapters.

Chapter 11
Separated Flow in Large Arteries

Abstract The main vascular pathologies in large arteries involve the presence of boundary layer separation, which gives rise to local fluid dynamic phenomena that escape from description based on simple models. Arteriosclerosis develops more frequently in regions that present anomalous spatiotemporal distribution of wall shear stress that is typically imputable to boundary layer separation. On the basis of this, the sites at higher risk of atherosclerosis are discussed. Then the fluid dynamics associated with the presence of stenosis, a local narrowing of a vessel due to arteriosclerosis, is described. The modifications following the different therapeutic options in the carotid and coronary arteries are discussed. Similarly, blood flow patterns that develop in the different types of aneurysms, a local enlargement of the vessel due to wall weakening, are described. They are discussed in relation to aneurism evolution and to therapeutic solutions.

11.1 Arteriosclerosis and Boundary Layer Separation

Arteriosclerosis is the deposition of substances transported with blood on the internal walls of the arteries provoking a progressive reduction of their lumen. The initial phase of arteriosclerosis can be imputed to multiple causes, like the inflammation of the arterial wall giving a thickening of the intima-media layer, pathologies of the endothelium reducing its protective function, or just the progressive deposition of lipid material. The individual arteriosclerotic risk level depends on numerous causes ranging from the properties of substances transported, their affinity with endothelium, to genetic predisposition. On top of all these biological reasons, certain characteristics of fluid dynamics play a fundamental role, both for the initiation and for the progression of arteriosclerosis, and represent a recognized risk factor for its development (Caro et al. 1969; Cunningham and Gotlieb 2005).

Flow and surrounding tissues can only interact through the exchange of dynamic actions: forces and stresses. Blood flow interacts with the endothelial layer of the arteries through the wall shear stress (WSS), which is recognized to have a primary role in the development of arteriosclerosis. On the biological side, an abnormal WSS on the endothelium triggers signaling that induces vascular inflammation (Chen et al.

© The Author(s), under exclusive license to Springer Nature Switzerland AG 2022   173
G. Pedrizzetti, *Fluid Mechanics for Cardiovascular Engineering*,
https://doi.org/10.1007/978-3-030-85943-5_11

2019). On the purely mechanical side, the endothelium is made of elongated cells that are kept aligned with the flow by the normal wall shear stress. When the wall shear stress is abnormal and not directed along the vessel, stresses may progressively alter this alignment of endothelial cells that get randomly oriented. In that case, the endothelium becomes rougher and more prone to deposition of lipid substances transported with the blood.

Wall shear stress changes during the heartbeat accordingly to the pulsatile nature of blood flow. Therefore, several measures were introduced to identify a relation between the presence of anomalous wall shear stress and arteriosclerosis. The most immediate measure is the value of the time-averaged wall shear (TAWSS) stress during the heartbeat, of duration T,

$$\text{TAWSS} = \frac{1}{T} \int_0^T \text{WSS} dt. \tag{11.1}$$

Low or negative values of TAWSS were shown to often correspond with the locations developing atherosclerosis. More modern indices were also introduced to better underline the importance of reversal of WSS for pathology (Lee et al. 2009); one of those is the oscillating shear index (OSI) that is defined

$$\text{OSI} = 1 - \frac{\left| \int_0^T \text{WSS} dt \right|}{\int_0^T |\text{WSS}| \, dt}, \tag{11.2}$$

which approaches zero when the WSS is predominantly positive and increases towards 1 when WSS reverses its sign during long time intervals.

The WSS quantity contained in (11.1) and (11.2) refers to the stream-wise component. In general, the wall shear stress is a vector tangent to the endothelium, as shown in formula (10.10). Vector quantities are more difficult to be synthesized into simple indicators because they also account for directional alterations that can be related to complex physiological phenomena. For example, the presence of helical flow, discussed above in Sect. 9.2, may help to reduce the presence of stagnating regions and the development of atherosclerosis (Morbiducci et al. 2011); however, the description of arterial blood motion at such detail is difficult with current diagnostic tools and a satisfactory identification of the role of fluid dynamics is still out of reach of clinical practice.

The general rule is that the risk of atherosclerosis is related to the presence of anomalous wall shear stress on the endothelium, sharp fluctuations, and spatial gradients, especially when stresses are reversed with respect to the main flow direction. It is evident that boundary layer separation and vortex formation are key causes for the development of flow reversal and anomalous wall shear stress. Additionally, regions

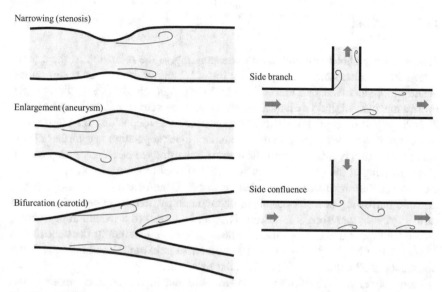

Fig. 11.1 Regions with higher chances of boundary layer separation, which are also higher risk of atherosclerosis

with flow reversal are associated with higher blood stagnation and material aggregation. Therefore, as a driving concept, the location of boundary layer separation can be confidently assumed as a region with a higher risk of atherosclerosis (Fig. 11.1).

It is fundamental to be aware of which regions may, at least qualitatively, present higher chances of developing boundary layer separation and thus higher risk of atherosclerotic developments. We have seen in the previous chapter that boundary layer separation occurs in those regions where velocity presents a spatial deceleration along the wall. Figure 11.1 displays some typical geometric conditions where this can happen. Separation is largely expected after a vessel narrowing/expansion, which can happen in presence of stenosis, at an enlargement, which is typically due to the wakening of the arterial wall leading to aneurysms. Both are pathological conditions that will be discussed in more detail later in this chapter. Boundary layer separation can also occur under physiological conditions for example at a bifurcation. In particular, the carotid bifurcation presents an enlargement (carotid sinus) on the side of the internal carotid, which represents a typical region at risk. In general, however, any branching leads to local flow decelerations that may give rise to boundary layer separation. We have also seen that a vortex, after its formation, interacts with the wall and gives rise to secondary boundary layer separation. Therefore, any important boundary layer separation may provoke secondary separation and create regions at risk even somehow away from those regions considered critical by geometric consideration only.

11.2   Stenosis

Stenosis is a pathological condition corresponding to the reduction of the arterial lumen due to atherosclerotic deposition; Fig. 11.2 shows the various stages during the development of stenosis that ends up with the progressive blockage of the vessel. From a mechanical, fluid dynamics perspective, it can start from a small disturbance in the flow that reduces the wall shear stress and facilitates deposition of material. This reduction of the vessel size provokes boundary layer separation and further disturbance that in turn reduces the wall shear stress downstream and enhances deposition. The narrowing of the vessel reduces the availability of blood-transported oxygen in the downstream organs served by that vessel; a phenomenon known as ischemia. Stenotic narrowing is a self-sustained phenomenon where its appearance induces further growth; therefore, it is extremely unlikely to record a reduction of stenosis during the time in absence of a therapeutic intervention. Eventually, the stenosis can even lead to a blockage of the vessel and give rise to extreme consequences to the organ supplied by the downstream vascular network.

Stenosis reduced the flow and causes ischemia to the regions whose blood (oxygen) allowance is provided by that vessel. Sometimes, secondary circulatory pathways can partially overcome this issue, although allowance through secondary vessels is rarely sufficient when the oxygen request increases above a minimum rate, because of exercise or stress.

In addition to the partial or total blockage of the vessel, the main risk of a stenosis is its partial breaking with the release of a small fragment (a thrombus) that is transported downstream. Along the branching arterial network, vessels become progressively smaller until such transported element is unable to pass through and, at a certain level of the branching, it gets blocked in a vessel closing it and not allowing blood availability to the tissues perfused by that vessel. For this reason, stenosis is also studied to assess its "vulnerability" to break up and release fragments. This depends on whether the stenosis is well perfused, it is hard passive material or it is composed of different blocks of materials. The process of rupture of a stenosis

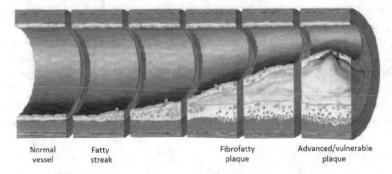

Normal   Fatty   Fibrofatty   Advanced/vulnerable
vessel   streak   plaque   plaque

Fig. 11.2 Progression of arteriosclerosis from normal vessel to formation of a critical stenosis (credit: Npatchett, CC BY-SA 4.0, via Wikimedia Commons)

can be influenced—besides its composition—by the presence and entity of vortex formation therein. Indeed, it was shown before, in Eq. (10.13), that vortex formation gives rise to dynamic hammering on the underlying tissue that may increase the risk of making the stenosis unstable and release fragments.

One typical site at risk of stenosis is the two carotid bifurcations (symmetric on the two sides of the neck) as shown in Fig. 11.3. The right or left common carotid artery bifurcates into the external carotid artery, bringing blood mainly to the muscular elements of the head, while the internal carotid artery grants blood allowance to the brain. The latter is of greater importance in relation to the associated pathological consequences. The internal carotid artery is also one of the vessels at higher risk of atherosclerosis because it originates at the carotid sinus, an enlargement right at the bifurcation that is a site where separation can occur naturally and is particularly prone to the development of stenosis. The consequence of stenosis at the internal carotid is the reduction of blood allowance to the brain. Its partial breaking can have consequences like ictus, which can be severe or temporary (TIA, transient ischemic attach) or be fatal leading to death.

Carotid sinus is so prone to the development of the atherosclerotic plaque that it is commonly monitored as an indicator for the individual predisposition to develop stenosis in other segments of the arterial network. Its analysis is also relatively easy

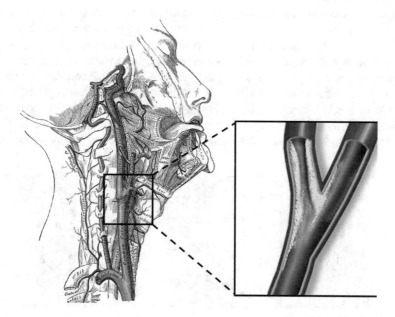

Fig. 11.3 Stenosis in the carotid artery (credit, left picture: Henry Vandyke Carter, Public domain, via Wikimedia Commons; right picture: Blausen.com staff (2014). "Medical gallery of Blausen Medical 2014". WikiJournal of Medicine 1 (2). https://doi.org/10.15347/wjm/2014.010. ISSN 2002–4436, CC BY 3.0, via Wikimedia Commons)

because the carotid is a superficial artery on the neck and can be visualized with good quality by simple ultrasound imaging like Doppler echography.

Therapeutic strategies for carotid stenosis at the early stage are made of blood thinners to reduce the risk of further aggregation and growth. When the stenosis presents a high degree of blockage or it is at risk of rupture, therapies are based on regular (invasive) surgery or endovascular (trans-catheter) surgery. Carotid endarterectomy is a common surgical approach to remove the arteriosclerotic plaque at or near the carotid bifurcation. Its diffusion also follows the relatively simple access to the carotid bifurcation. The procedure is schematically sketched in Fig. 11.4. The carotid lumen, once the blood transit is temporarily deviated, is accessed through a longitudinal cut on the arterial wall. The deposited material is then removed and the artery is sutured. During the suture, a small patch is commonly added to the artery wall to avoid a reduction of the lumen of the sutured artery. Evidently, the shape of such a patch influences the geometry of the reconstructed vessel, the distribution of wall shear stress, which in turn influences the risk of restenosis after surgery. Patches are made large enough to ensure a good passage of blood; however, they must not be too large to avoid excessive enlargements and boundary layer separation, which is a major risk factor for the therapeutic outcome. Monitoring the flow in the reconstructed artery, for example with color Doppler ultrasound, is important to assess the risk associated with fluid dynamics.

Invasive surgical therapies are often substituted by endovascular procedures; shown schematically in Fig. 11.5. The endovascular approach is commonly preferable to subjects with additional risk factors, like aged patients, which may suffer from an invasive procedure. In endovascular surgery, the vessel is accessed through a guided catheter that releases an endovascular prosthesis (stent). A balloon is expanded pressing the plaque at the wall, without removing it, and a prosthesis is placed on the expanded vessel restoring a sufficient lumen to allow blood passage. This prosthesis alters the vessel geometry and may create an elasticity mismatch. These changes affect the fluid dynamics and the interaction between flow and tissue, which may,

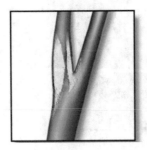

Fig. 11.4 Procedure of carotid endarterectomy surgery. The carotid presents a stenosis at the bifurcation (left image), the vessel is opened and atherosclerotic material is extracted (middle), the vessel is then sutured (right), (credit: Blausen.com staff (2014). "Medical gallery of Blausen Medical 2014". WikiJournal of Medicine 1 (2). https://doi.org/10.15347/wjm/2014.010. ISSN 2002–4436, CC BY 3.0, via Wikimedia Commons)

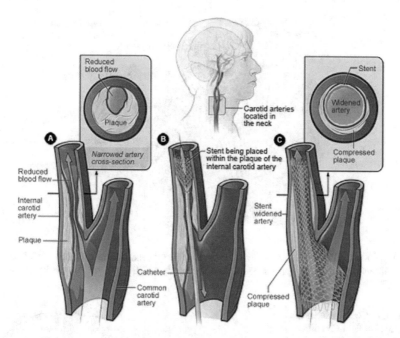

Fig. 11.5 Carotid endovascular prosthesis. Internal carotid artery with stenosis (**a**), prosthesis placement with catheter (**b**), final configuration with normal blood flow restored (**c**) (credit: National Heart Lung and Blood Institute, NIH; Public domain, via Wikimedia Commons)

in turn, alter the distribution of wall shear stress. Cases of restenosis are observed and may sometimes be imputable to the alteration of blood motion, which should be monitored as a measure of the quality of the therapy.

Another major site at risk of stenosis is the coronary tree. Coronaries are the arteries that bring oxygenated blood to the myocardium, the heart muscle. The two main coronaries, the right and left coronary arteries (RCA and LCA), originate just behind the aortic valve from two of the three sinuses of Valsalva (described later in Chap. 13). Thus, the heart pumps blood from the left ventricle cavity into the aorta and, right after the aortic valve, a part of that blood returns to the heart to feed its own myocardium. As shown in Fig. 11.6, the RCA feeds the myocardium on the side of the right ventricle, the LCA divides into circumflex and in the anterior and posterior interventricular arteries to feed the left ventricle and the interventricular septum.

We previously discussed carotid stenosis as a life-threatening disease because it reduces blood allowance to the brain. Similarly, coronary stenosis is a life-threatening disease because reduces blood allowance to the heart. The consequence of a coronary stenosis (see Fig. 11.7) is the ischemia of the myocardium that, for this reason, reduces its ability to contract. When the degree of coronary blockage is almost complete, the oxygen allowance reduces to near zero and the region of muscle perfused by that coronary can undergo myocardial infarction. When the lack of oxygen persists for some time that tissue dies and becomes necrotic.

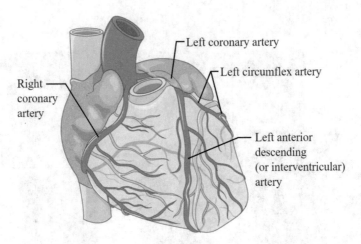

Fig. 11.6 Major coronary arteries that supply blood to the myocardium (credit: Servier Medical Art. Annotations by Mikael Häggström, M.D. Reusing images, CC BY 3.0, via Wikimedia Commons, adapted)

Fig. 11.7 Coronary stenosis and myocardial infarction (credit: Blausen Medical Communications, Inc., CC BY 3.0, via Wikimedia Commons, adapted)

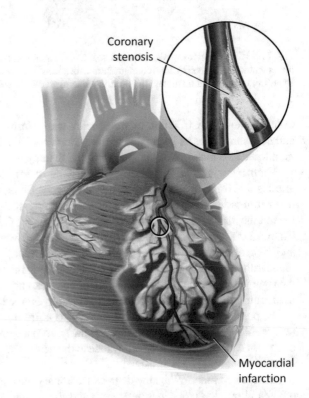

Myocardial ischemia, or infarction, affects the ability of the heart muscle to contract, thus the ability of the heart to pump blood into circulation. An extended infarction, due to stenosis in a large upstream artery, leads to the inability of the heart to pump enough blood and can lead to death if not recovered rapidly before the infarcted tissue dies. A small infarction or ischemia is primarily detected in terms of reduction of the cardiac function; thus, their symptoms are those of a cardiac disease, and they are first detected by cardiac dysfunction, although they originate from a vascular disease.

The most common approach to recognize the presence of myocardial ischemia is thus that of echocardiography to observed whether some regions of the cardiac wall present reduced contraction. Sometimes, the blood allowance is sufficient for an approximately normal contraction at rest, while it becomes insufficient under stress or exercise. Therefore, it is also common to perform a stress echocardiography (by exercise, or using pharmacologic stress in patients who cannot perform exercise) to recognize contraction abnormality in presence of a higher demand for oxygen. This occurs in the presence of small stenosis as well as when some blood is able to reach the region through secondary circulatory pathways. Suspected coronary stenosis is then verified by coronary angiography that permits to visualize the blood flowing into the coronary tree and thus the lumens of the coronary arteries. Such assessments are based on geometric properties only and do not consider explicitly the effective fluid dynamic performance of the vessel presenting a stenosis. This can lead to an overestimation or an underestimation of the narrowing in terms of effective functional relevance.

A more advanced approach to evaluate the impact of a stenosis is based on measuring the fractional flow reserve (FFR), the ratio between blood pressure upstream and downstream the narrowing to verify whether there can be secondary pathways ensuring blood allowance downstream the lesion. Coronary pressure measurements for FFR are performed during coronary angiography with a coronary pressure guidewire (Pijls et al. 1996). FFR is considered the most reliable assessment of coronary artery disease; however, the need for invasive measurements may limit its routine diagnostic application. Therefore, noninvasive alternatives to FFR were later introduced based on reproducing the fluid dynamics in the coronary vessel by mean of numerical solution of the Navier–Stokes equation in the geometry extracted by CT angiography (Taylor et al. 2013). This approach, which took the name FFR_{CT}, was successfully evaluated in a series of clinical trials and it is currently proposed as a viable clinical option. On the other hand, this is a purely virtual analysis and the reliability of results depends on the accuracy of the geometry extracted from CT and of the flow conditions inserted at the initial and terminal boundaries of the simulated coronary tree. The rapid advancement of imaging technology led to the image-based alternative to FFR based on measuring properties of blood flow across the stenotic constriction using a sequence of images in coronary CT. Such an approach, usually called quantitative flow reserve (QFR), has the advantage of being based on in vivo data; it showed good performance in comparison to invasive FFR and appears a viable alternative for routine clinical application (Tanigaki et al. 2019).

Early therapy for coronary artery disease is one of the blood thinners to avoid their progression; when the degree of the disease exceeds given limits, resolutive intervention has to be planned. The open-chest surgical approach is that of by-passing the blocked vessel with the insertion of a new artificial vessel, a by-pass graft, that connects the vessel upstream the stenosis with the vessel downstream. At the position of the junctions where the graft connects with the original coronary, blood flow can be disturbed and that region can become at further risk of stenosis. Much more common, however, is now the use of endovascular procedures. The procedure is shown schematically in Fig. 11.8: the vessel is reached by a guided catheter from the aorta. The catheter is equipped with a balloon that expands the endovascular prosthesis and a prosthesis (stent) is placed and remains in position after the catheter is released. The changes in geometry and elasticity about the stent position may sometimes disturb the fluid dynamics and alter the distribution of wall shear stress. However, these are subjects at risk, where cases of restenosis can be as frequent as those of new stenosis, and are commonly kept under periodic control.

Carotid and coronary are the most common sites at risk of stenosis; they take special relevance because they supply oxygen to life-supporting organs like the brain and the heart, respectively. However, stenosis can develop in numerous other sites along the arterial tree. Common examples are the side branches on the aorta or the iliac bifurcation at its end. Nowadays, most arterial stenosis diseases can be treated by endovascular procedures, whose technology is continuously advancing. Stents are available for about any dimension and shape, and multiple stents can also be combined to reconstruct bifurcations and multiple branching. Typically, patients who developed stenosis are subjects with a higher predisposition to atherosclerosis. Therefore, alteration of the fluid dynamics in such patients must be carefully monitored in those sites where boundary layer separation is likely or is observed.

11.3 Aneurism

Aneurism is a pathology due to the enlargement of the vessel over a short tract. The tissue stretched in such expansion is typically thinner and weaker and the main threat is associated with the risk of rupture. Aneurysms are more frequent in the aorta, at all levels, and in the brain arteries. The main issue associated with aneurysms is that in most cases they do not give flow impairment and do not produce symptoms. Therefore, they can be detected after specific searches indicated for inheritance or other risk factors, frequently they are detected simply by chance. Aneurism is a silent disease: despite its common absence symptoms, when the aneurysm undergoes rupture it can be fatal.

Schematically, aneurysms are divided into two main geometric types as shown in Fig. 11.9. A fusiform aneurysm is a dilatation of the entire vessel that is characterized by a diameter larger than normal. Differently, a saccular aneurysm is a side bulging of the vessel tissue, which generates a balloon-like protrusion. Evidently,

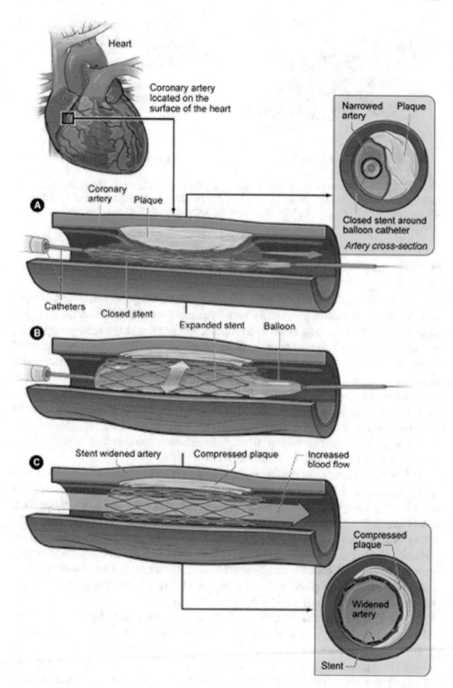

Fig. 11.8 Endovascular coronary surgery artery (credit: National Institutes of Health, Public domain, via Wikimedia Commons)

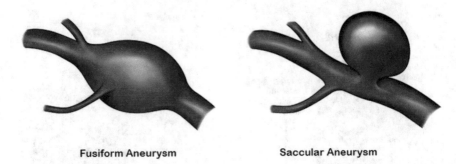

Fusiform Aneurysm   **Saccular Aneurysm**

Fig. 11.9 Type of aneurysms (credit: Withers, K., Carolan-Rees, G. & Dale, M. Pipeline™ Embolization Device for the Treatment of Complex Intracranial Aneurysms. Appl Health Econ Health Policy 11, 5–13 (2013). https://doi.org/10.1007/s40258-012-0005-x. Creative Commons Attribution License)

the categorization is not always so sharp and numerous intermediate conditions may also exist.

The fluid dynamics inside an aneurysm depends on the geometrical details of its specific shape. Fusiform geometries usually present a central jet due to boundary layer separation at the expansion and recirculating regions at the enlargement. The jet may or may not be aligned with the distal vessel and possibly impact one side wall of the aneurysm. In a saccular geometry, the flow is mainly stagnating therein with more or less wash-out of the blood. Therefore, the first fluid dynamics phenomenon in aneurysms is the presence of stagnation areas, which may form thrombi when there is not enough exchange of blood with the main flow. The second important phenomenon is the impact of the jet on the side wall provoking overpressure in the splash area; an impact occurring at every heartbeat hammering on a wall that is already thin and weak may increase its risk of rupture.

The birth of aneurysms can be imputable to the local weakening of the tissue. This phenomenon is sometimes related to alteration of the local fluid mechanics that creates overpressure or wearing shear stress at the wall. More frequently, however, this is imputable to an alteration of the tissue itself for multiple causes that often follow genetic predisposition. The progression and development of the aneurysm are primarily due to the continuing presence of the causes that generated it. Progression, however, can also be imputable to the specific alteration of the fluid dynamics therein. The major risk is its rupture that can bring to ictus (brain aneurysm) and internal hemorrhage, which in turn can lead to sudden death.

A typical site for the development of aneurysm is the aorta. An excessive wall deformation in the proximal part of the ascending aorta can follow from genetic causes or also be a consequence of anomalies in the aortic valve. In the latter case, the valve jet may present high velocities that are deviated toward the aortic wall, because the orifice area is small and tilted. The impact of the jet on the wall creates high shear that can wear down the epithelium and also produces a continuous hammering due to over-pressure acting on such wall. This type of aneurysm is sometimes associated with the presence of valvular stenosis or bi-leaflet aortic valves, discussed later in Chap. 13, whose opening may provoke a laterally directed jet with high velocity. In the aortic arch, including part of the ascending and descending aorta, the aneurysm develops mainly because of genetic alteration of the wall tissue. The abdominal aorta above the iliac bifurcation is one of the most frequent sites for the formation of aneurysms, which deserved its own acronym AAA standing for Abdominal Aortic Aneurysm.

Fluid dynamics plays a role in the progression of the aneurysm. Consider a saccular aneurysm first. The flow may occur mostly along the vessel, without significant exchange with the side expansion; for example, when the bulge is very lateral and the opening is small and aligned with the vessel wall as sketched in the left side of Fig. 11.10. In this case, blood coagulation can likely develop inside the aneurysm and remains therein to possibly protect the bulged wall. This aneurysm is stable, from a fluid dynamic point of view, because the flow is not expected to induce its growth. On the opposite, when the main flow partly enters into the side bulge, as shown on the right sketch of Fig. 11.10, it can provoke additional shear and epithelial damage, it does not allow coagulation of blood, thus keeps the bulged camera active. In this case, the aneurysm is unstable, by a fluid dynamic perspective, because it is expected to progress and to present an increased risk of rupture. Similar evaluations can be brought forward about fusiform aneurisms. In these cases, complete coagulation is less common. Examples of flow in active aneurisms are shown in Fig. 11.11, the deviation of the main flow may induce an increase of pressure on a wall that can affect the aneurysm progression.

Fluid dynamics, however, is rarely used clinically to categorize the risk of progression or rupture of the aneurysm. Currently, risk assessment is essentially based on the size of expansion only. However, the progression of imaging techniques now allows the evaluation of the intra-aneurysm blood velocity vector field and novel solutions are under development to improve the categorization and support diagnosis and therapeutic planning (Chung and Cebral 2015).

Once an aneurysm has been detected, there are no specific pharmacological therapeutic treatments (besides those for associated risk factors, like high blood pressure). The periodic control is crucial to monitor its progression. Therapies are essentially of surgical or endo-surgical type as sketched in Fig. 11.12. Surgery is performed

Fig. 11.10 Flow in a saccular aneurysm where the aneurysm is separated from main flow (left) or when it exchanges blood and provokes shear inside the aneurysm

Fig. 11.11 Flow in an active saccular aneurysm (left) and in a fusiform aneurysm proximal to the iliac bifurcation (right) (credit: Tezduyar et al. Int. J. Numer. Meth. Fluids 2008;57:601, with permission from John Wiley and Sons)

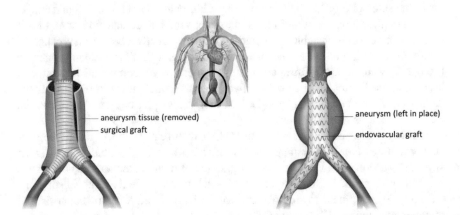

Fig. 11.12 Surgical (left) and endovascular surgical (right) treatment of an abdominal aortic aneurism (credit, left and right pictures: Wanhainen et al. Eur. J. Vasc. Endovasc. Surg. 2019;57:8, with permission from Elsevier; middle image: Blausen.com staff (2014). "Medical gallery of Blausen Medical 2014". WikiJournal of Medicine 1 (2). https://doi.org/10.15347/wjm/2014.010. ISSN 2002–4436., CC BY 3.0, via Wikimedia Commons)

through a bandage or, more likely, by removing the aneurysm and replacing the portion of the vessel with a prosthesis. More recently, endovascular surgery, which is performed by inserting a stent in the vessel is becoming a much more common option. It is important to remark that the prosthesis has to be sutured (for regular surgery) or anchored (endovascular surgery) in the vessel upstream and downstream of the region where the aneurysm was present. This sometimes represents a challenge to surgeons because the tissue was weaker and not intact at the aneurysm, it is, therefore, possible that some weakening is present in the nearby tissue that may not allow an optimal anchoring. It is important to verify after surgery that there is no leakage across the prosthesis and that there is not blood motion in the aneurysm

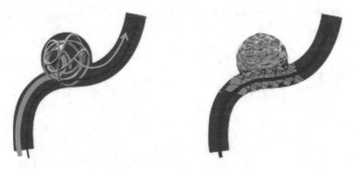

Fig. 11.13 Changes in the fluid dynamics from before (left) to after (right) endovascular surgery of a saccular aneurysm

region outside the endovascular prosthesis After the insertion of the endovascular prosthesis the blood should flow through the prosthesis while stagnating blood is left to coagulate in the lateral expansions that are excluded from the circulation as sketched in Fig. 11.13.

We said above that the causes of aneurysm formation are genetic or due to regional alteration of either tissue or flow properties. The surgical repair solves the effects but does not remove the causes that led to aneurism development. Therefore, frequent controls are important after surgical therapy close to the repaired vessel where tissues can have sub-optimal mechanical properties, as well as in other sites at risk. Monitoring is mainly performed looking at the vessel geometry; however, it is also important to verify the presence of anomalies in the flowing blood that witnesses abnormal dynamics and possibly associated risk factors (Ziegler et al. 2019).

Chapter 12
Cardiac Mechanics I: Fluid Dynamics in the Cardiac Chambers

Abstract This chapter starts by introducing the general electromechanical cycle of the beating heart. It sketches out the connection between the electric phases of the electrocardiogram and the mechanical phases of the left ventricle volume changes corresponding to active myocardial contraction and passive relaxation. Then the classical description of ventricular mechanical function based on the pressure–volume relationship is introduced posing attention to its connection with blood motion. The fundamental knowledge about the fluid dynamics inside the left ventricle is summarized, the main descriptors are divided into those describing the efficiency of blood transit and those regarding the exchange of force between blood and tissue. A categorized description of the principal ventricular pathologies is reported with a description of the associated alteration in cardiac fluid dynamics. The potential relationship between cardiac flow and heart failure is carefully discussed.

12.1 Cardiac Electro-Mechanical Cycle

The complete heart is an organ that contains two biological pumping systems, the right heart and the left heart. Each individual heart is composed of an atrium that receives low-pressure blood and is connected to the respective ventricle that pumps blood at the arterial pressure into the respective circulation. The left and right sides work synergistically in the whole heart and they present similar timing of their activity. They are also arranged in series along the circulatory network. The left heart pumps oxygenated blood in the primary circulation that, after releasing oxygen to all body, terminates into the right side of the heart. The right heart pumps de-oxygenated blood in the pulmonary circulation, where it entrains new oxygen and terminates in the left heart. Therefore, for blood incompressibility, each side pumps the same amount of blood volume in the circulation. The main difference is that the pumping work is performed at a significantly different arterial pressure; on the right side, pressure in the pulmonary artery typically ranges between 5 and 20 mmHg whereas on the left side pressure in aorta is much higher and varies normally from 80 to 120 mmHg.

© The Author(s), under exclusive license to Springer Nature Switzerland AG 2022
G. Pedrizzetti, *Fluid Mechanics for Cardiovascular Engineering*,
https://doi.org/10.1007/978-3-030-85943-5_12

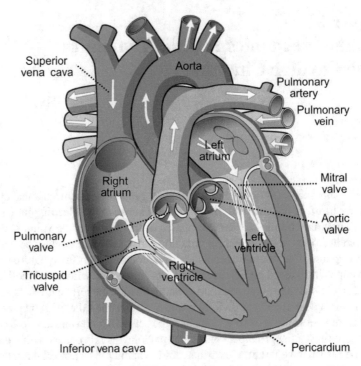

Fig. 12.1 Heart anatomy and blood flow paths (credit: Wapcaplet, CC BY-SA 3.0, via Wikimedia Commons)

The heart anatomy, with an indication of the blood flow pattern, is shown in Fig. 12.1. On the left side, the pulmonary veins coming from the lungs bring oxygenated blood to the left atrium. The left atrium connects to the left ventricle through the mitral valve, a valve with two leaflets (bicuspid valve) that opens into the left ventricle and avoids backflow. To this aim, and due to the high-pressure difference that can develop between the left ventricle and left atrium, leaflets are retained from opening into the atrium by the chordae tendineae that connect the tips of the valvular leaflets to the inside of the ventricle wall in a reinforced region called papillary muscles of the myocardium. The myocardium is a thick muscle that surrounds the left ventricle and permits its contraction to vigorously pump blood into the aorta, the first artery of the primary circulation, and work against the high aortic pressure. The aortic valve is placed at the base of the ventricle, on the right side of the mitral valve, and separates the left ventricle from the aortic artery. It is a tricuspid valve (with three leaflets) that avoids backflow, for the relatively lower pressure difference from the aorta to the left ventricle, by the closure of the three leaflets with the tips aligned downstream. On the right side, the right atrium receives poorly oxygenated blood from the inferior and superior venae cavae and connects to the right ventricle through the tricuspid valve. The right ventricle is surrounded by a thin myocardial layer and pushes blood through the pulmonary valve, into the pulmonary artery. The right ventricle produces the same volume rate as the left, but it works against much the lower pressure in the pulmonary arteries, for this reason, the myocardium that surrounds the ventricle on the right side is thinner than that on the left side. Geometrically, the left ventricle has roughly the shape of a prolate spheroid and the right ventricle wraps around it on the right side, for about 45°, in a triangular shape. The two ventricles are separated by a portion of a thick myocardium called the interventricular septum.

The left ventricle (LV) is the principal mechanical element of the human heart. It has the function of a volumetric pump that receives low-pressure blood from the venous system through the left atrium and ejects it with higher pressure through the aortic valve into the primary circulation. The thick muscular layer, the myocardium, that surrounds the LV chamber operates in a sequence of mostly passive relaxations when it receives the blood, and active contractions to push it into the circulation. Given the fundamental mechanical function of the heart, the myocardial tissue deformation and the blood flow inside the LV represent a central issue of clinical evaluations.

LV function can be described, in global terms, as comprised of interconnected activities; the electric stimulation of muscular fibers (*electric cycle*), which gives rise to changes in the ventricular volume (*mechanical cycle*) that, in turn, develop in presence of a ventricular pressure and produce *mechanical work*. Cardiac electric cycle develops by the propagation of the electric signal that produces the mechanical contraction of the individual myocardial cells and eventually gives rise to the volumetric reduction of the chambers. For this reason, the cardiac cycle is commonly referred to as an *electro-mechanical cycle*. The electrocardiogram (ECG) records the polarization and de-polarization of the muscular fibers, due to electrical voltage differences, which corresponds to the beginning of fibers contraction and relaxation, respectively. One typical ECG trace is reported in Fig. 12.2 next to a sketch of the main electric conduction system. The electrical stimulation starts from the sinoatrial node placed about the tip of the atrium (on the right side) and propagates into the myocardium surrounding the two atria. It produces polarization and consequent

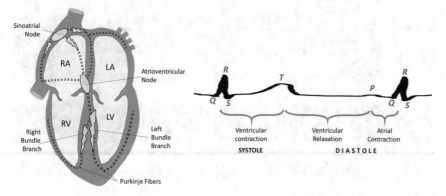

Fig. 12.2 The electric cycle. Left: fibers transmitting the electric impulse, right: typical electrocardiogram (ECG) signal

shortening of the muscular fibers: the *atrial contraction*; which pushed some blood into the ventricle; this weak polarization is noticeable in the ECG by a small peak that is called the P-wave. The electrical conduction converges into the atrioventricular node, placed between the ventricles and the atria where it slows down before propagating rapidly into the ventricles' branches. The QRS complex in the ECG indicates the polarization of the ventricular myocardial fibers, after which the ventricular contraction develops. The ventricular contraction, or *systole*, pushes blood into the circulation. When the contraction is completed the muscular fibers depolarize, as revealed by the T-wave in the ECG, and relax allowing the blood to fill the ventricle during *diastole*. Diastole is then completed by the following atrial contraction.

The electric cycle has a parallel mechanical cycle comprising ventricular filling and ejection to form the electrical–mechanical cycle. We will keep the focus on the LV, unless otherwise specified, that is the most energetic element of the human heart; however, the right ventricle follows in parallel an analogous process. With reference to Fig. 12.3, we can correlate the electric cycle with the mechanical events. During systole, the LV contracts, the mitral valve is closed, its volume decreases, and flow is ejected (S-wave) at systolic pressure through the aorta; during deceleration of the S-wave pressure in the LV decreases and falls below that of the aorta until the aortic

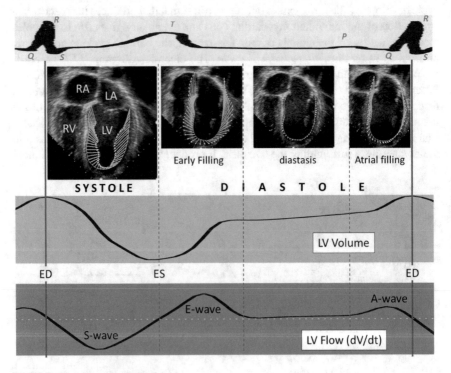

Fig. 12.3 The electromechanical cycle

valve closes. Then the myocardium relaxes, and changes its shape for the rapid de-activation of the muscular fibers, during this short period with both valves closed (called iso-volumic relaxation) pressure rapidly decreases until it falls below that in the left atrium, the mitral valve opens and blood flows into the LV. This is the early filling phase of diastole, the E-wave, that corresponds to the main increase of LV volume. This phase terminates when the atrial and ventricular pressure becomes comparable and flow is very small reaching a condition of diastolic stasis, called diastasis. Afterward, the atrial contraction completes the LV filling (A-wave) and the diastolic phase. The mitral valve closes and, after a short iso-volumic contraction phase, the aorta opens and systole restarts.

The volumetric function of the LV is primarily described through parameters as those of a volumetric pump. The volume at end-diastole, V_{ED}, is the maximum size of the LV chamber that then contracts to reach a minimum volume value at end-systole, V_{ES}. Therefore, the stroke volume $SV = V_{ED} - V_{ES}$ is the volume of blood ejected by the LV into the circulation, as well as the volume entering during diastole. The SV is also the volume that passes through any cross-section of the circulatory network during one heartbeat.

The SV is commonly normalized with the V_{ED} to provide a dimensionless measure of the entity of the contraction relative to the available volume. This measure is defined as *ejection fraction*

$$EF = \frac{V_{ED} - V_{ES}}{V_{ED}} = \frac{SV}{V_{ED}}, \qquad (12.1)$$

which represents the most common clinical parameter to assess the LV systolic function. Evaluation of EF requires the evaluation of LV volumes, which can be performed with numerous methods based on imaging, from echocardiography to MRI, for example. In normal hearts, the EF is usually about 65%, or above and it is considered abnormal when it falls below 55% (although exact figures depend on the measurement method). A reduction of the EF reveals the presence of a cardiac dysfunction, although there are also pathologies that develop in presence of a preserved EF.

A deeper understanding of LV mechanical function must include the role of pressure to identify the effective *mechanical work* associated with the electromechanical cycle. Pressure inside the left ventricle reaches its maximum value during the systolic contraction, when it has to overcome the systolic pressure inside the aortic artery (normally about 120 mmHg) and reduces during the contraction to values comparable to the diastolic pressure in the aorta (about 80 mmHg). After the closure of the aortic valve, and opening of the mitral valve, the pressure inside the ventricle during diastole falls to the low value that is found in the left atrium, reaching a minimum of a few mmHg. The interplay between pressure and volume led to an interpretation in terms of the isothermal thermodynamic process that can represent a pressure–volume loop as sketched in Fig. 12.4 for a normal subject. The loop is bounded above and below by the end-systolic pressure–volume relationship (ESPVR) that contains the

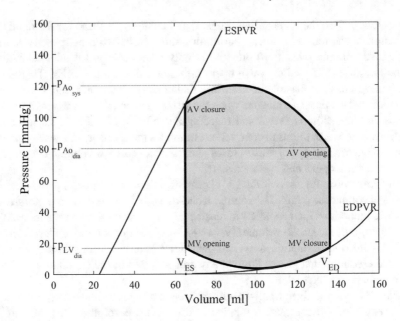

Fig. 12.4 Pressure–volume loop in a normal left ventricle

end-systolic point, and its end-diastolic counterparts (EDPVR). These are related to the concept of elastance that is beyond the scope of this book, more details can be found in classic books on cardiac physiology (Berne and Levy 1997).

The pressure–volume loop must be run in the counter-clockwise direction. The upper curve represents the systolic contraction when volume reduces from end-diastole to end-systole with pressure reaching its maximum about half the way. The lower curve corresponds to the diastolic LV expansion. The approximately vertical lines connecting the two curves represent the iso-volumic phases corresponding to the quick transition between the closure of one valve and the opening of the other when the volume is approximately constant and pressure rapidly changes.

The pressure–volume loop is relevant for the evaluation of the mechanical work of the ventricle. The instantaneous power associated with the volume of fluid, which represents the time derivative of the mechanical work, is defined as the scalar product between acting force and velocity. In the case of a volume of fluid $V(t)$ with no volumetric forces, the only force is the pressure that acts normally to its bounding surface $S(t)$, and the total power is

$$\mathcal{P}(t) = \int_S p\boldsymbol{v} \cdot \boldsymbol{n} dS. \tag{12.2}$$

Pressure can be confidently assumed as approximately constant in the chamber with a value $p(t)$; in this case, it can be taken out of the integral and the integral is identically zero in an incompressible fluid for mass conservation (4.3). This result is informative when one considers that the wall is given by the tissue boundary and the open parts where fluid flows $S = S_b + S_{open}$. Equation (12.2) can be rewritten as

$$p \int_{S_b} \mathbf{v}_b \cdot \mathbf{n} dS + \int_{S_{open}} \mathbf{v}_f \cdot \mathbf{n} dS = 0;$$

and including the boundary of the open part

$$p \int_{S} \mathbf{v}_b \cdot \mathbf{n} dS = -p \int_{S_{open}} \left(\mathbf{v}_f - \mathbf{v}_b \right) \cdot \mathbf{n} dS.$$

The first integral represents the volumetric rate and Eq. (12.2) becomes

$$p \frac{dV}{dt} = -p \int_{S_{open}} \mathbf{v}_{f_r} \cdot \mathbf{n} dS, \qquad (12.3)$$

where $\mathbf{v}_{f_r} = \mathbf{v}_f - \mathbf{v}_b$ is the fluid velocity relative to the (possibly moving) valvular boundary. Equation (12.3) shows that the instantaneous power required by the myocardium to change the volume against the pressure is equal to the power of the fluid leaving the LV cavity against the same pressure. It further tells the total mechanical work performed by the LV during one heartbeat, which is the time integral of either terms in (12.3), is the area of the pressure–volume loop

$$W = \int_0^T p(t) \frac{dV}{dt} dt = \oint p(t) dV. \qquad (12.4)$$

Looking at Fig. 12.4, it is evident that most of the mechanical work is performed during systole, and can be estimated as

$$W_{sys} = \int_{V_{ES}}^{V_{ED}} p \, dV \cong \bar{p}_{sys} SV, \qquad (12.5)$$

where $\bar{p}_{sys}$ is the average value of aortic pressure during systole.

The properties described so far provide information about the overall mechanical performance of the LV associated with the exchange of a volume of fluid. However, these are only global evaluations that do not account for the details of fluid dynamics that develop inside the left ventricle and that influence the efficiency of its mechanical function.

12.2   Fluid Dynamics Inside the Left Ventricle
with a Mention to the Other Chambers

Heart function is about creating and sustaining the motion of blood. The previously discussed electromechanical cycle has, therefore, its ultimate effect in the dynamics of blood flowing through the LV from the mitral to the aortic valve.

Despite this apparent simplicity of the heart cycle, the fluid dynamics inside the left ventricle is a very intense dynamical phenomenon and represents a fundamental element in cardiac function. The incoming jet that enters the LV develops impulsively; within a few hundreds of seconds, it reaches speeds above the meter per second to enter a few centimeters long cavity. Then, just as rapidly, the flow must reverse the direction of motion of 180° and redirects toward the aorta where it will exit at a similarly high speed. The diastolic jet develops boundary layer separation from the tips of the mitral valve and immediately gives rise to a swirling motion within the cavity, as exemplified in Fig. 12.5 (left picture). The mitral orifice is slightly offset with respect to the ideal ventricular axis for which the jet redirects toward the lateral wall and gives rise to an asymmetrical swirling structure. The underlying phenomenon is that of the formation of an irregularly shaped vortex ring, both during the E-wave and during the A-wave, which then dissipates and stretches toward the outflow tract at the beginning of systole as shown in Fig. 12.5 (right picture).

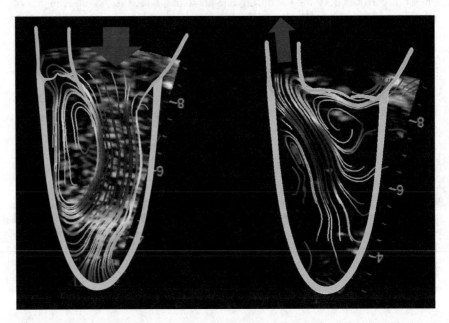

Fig. 12.5 Blood motion inside the left ventricle, sketch superimposed on the streamlines on the central longitudinal plane reconstructed from echocardiography. Image during diastolic filling (left) and systolic ejection (right)

Fig. 12.6
Three-dimensional vortex
structure inside the left
ventricle during late diastolic
filling computed from
numerical simulations. The
ring-shaped vortex created
during early filling is inside
the ventricle and a newly
generated ring is forming
during atrial contraction

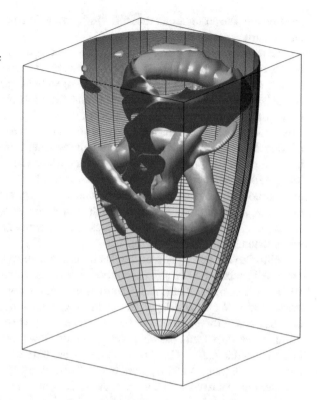

The flow pattern in a normal LV was extensively described in the literature (Kilner et al. 2000; Pedrizzetti and Domenichini 2005). It can be qualitatively understood in more depth in terms of three-dimensional vortex dynamics. In brief, and with reference to Fig. 12.6, the inflow jet during the diastolic E-wave enters through the mitral valve and develops a ring-like vortex structure below the valve, which represents the jet head. This vortex ring is slight displaced with respect to the chamber because the mitral valve is not central, and it slows down on the side closer to the boundary (the posterior-lateral wall, behind the shorter mitral valve leaflet) for irrotational image effect. The ring thus tilts with one side remaining more upstream while the other reaches the center of the chamber; shortly after, the vortex induces boundary layer separation on the same side and dissipation is enhanced therein. Eventually, the vorticity remains stronger about the center of the chamber where a rotatory motion develops. The same phenomenon develops, in a weaker form, during the A-wave, which feeds the previously developed dynamics. Sometimes, in healthy subjects and during exercise, this process is intense, the Reynolds number is high enough that it can be accompanied by vortex instabilities giving rise to weakly turbulent flow. Typically, at the end of diastole, the blood presents a weak rotation with velocity directed upward toward the LV outflow tract. This facilitates the blood ejection into the aorta during the following contraction and avoids sharp dissipation during early systolic ejection.

The length of the jet, the phenomena associated with its impact on the endocardial tissue, as well as the development and dynamics of the vortex structure inside the cardiac chamber, depend on various physiological and pathophysiological factors. A fundamental role is given by the size and the geometry of the chamber and its synergistic contraction and elastic relaxation, as well as the geometry of the mitral valve orifice. All these concurring elements can alter the picture discussed above, making the vortex a stable structure maintaining kinetic energy or an unstable structure that creates turbulence. It must also be considered that blood is an incompressible medium. All myocardial regions must work in harmonic synergy to push blood toward the aortic exit and receive blood evenly; an incorrect timing of contraction or relaxation in one region of the wall has the result of pushing blood toward the other region, thus creating intraventricular pressure gradients that stress the facing tissue rather than creating blood motion.

Following the description of the featuring phenomena of LV fluid dynamics, we can briefly mention what is known about the other chambers (Kilner et al. 2000). Blood motion in the left atrium is driven by the pulmonary veins that enter the atrium transversally. Very much dependent on the angle attaching these veins, the resulting flow can take a rotary motion or be more irregular and weakly turbulent; when the mitral valve opens this possibly rotary motion flows down into the left ventricle in a funnel-like patter. These considerations are extracted from few visualizations and not much more is known (Park et al. 2013). Similarly, little is known about the right atrium that receives blood from the inferior and superior cavae veins, they are supposed to produce a rotary motion and can give rise to turbulence depending on the orientation of the two jets.

The right ventricle (RV) presents a peculiar shape that is elusive to visualization methods and limits the possibility to assess its mechanical function. It is characterized by a streamlined geometry, with a relatively wider region below the tricuspid valve that narrows as the chamber extends around the LV and converges toward the pulmonary valve. However, the significance of this geometry on RV fluid dynamics and function is largely unknown. Currently, there are only a few works performed with respect to the characterization of RV fluid dynamics, although RV function has been shown to be a major determinant of clinical outcome in numerous cardiac dysfunctions including congenital heart diseases. Additionally, it is worth mentioning that RV function is not marginal as the volume of blood ejected from the RV must be equal to that ejected by the LV because the circulation system is a closed one. For the same reason, a reduction in the function of one ventricle immediately reflects on the other.

The flow field in the RV presents some qualitative differences with respect to the LV due to its specific geometry (Fredriksson et al. 2011; Mangual et al. 2012; Collia et al. 2021). The diastolic filling presents some analogy to that in the LV with the development of a ring-like vortex structure behind the tricuspid valve during the inflow; this forming vortex interacts with the close boundaries of this slimmer chamber, particularly on the septal side. The vorticity remains mainly in the region from below the valve to the apex and does not spread much in the region toward the outflow. During systolic contraction, the remaining vorticity extends along the converging outflow tract adding a slightly helical pattern to the otherwise largely irrotational velocity field.

12.3   Evaluation of LV Fluid Dynamics

The role of fluid dynamics inside the LV (as well as in the other chambers) can be categorized into two fundamental aspects. The first is a kinematic aspect, about the efficiency of the flow transit; the second is a dynamic aspect, about the exchange of forces between flowing blood and surrounding tissues and energetic efficiency (Pedrizzetti and Domenichini 2015).

(i)   *Kinematic aspects: Flow transit*

The quality of flow transit corresponds to mapping the time of residence of blood inside the chamber and the formation of regions with a reduced exchange of blood. The reduction of the wash-out of blood in the LV, as may occur in presence of stagnation regions, represents a risk factor for thrombus formation; especially when the higher residence time is accompanied by a high shear stress that can trigger platelet activation and aggregation mechanisms.

Major advancements about this aspect were achieved by processing 3D phase-contrast MRI acquisitions, usually called 4D Flow MRI, that provides the 3D velocity vector field in the entire LV (with the limitation of moderate time and space resolution, and of reconstructing the phase-averaged flow field from numerous heartbeats). There, flow transit was analyzed through the subdivision of the LV end-diastolic volume of blood, V_{ED}, into 4 sub-volumes depending on whether the corresponding blood resides more or less than one heartbeat in the LV chamber (Bolger et al. 2007). The division is as follows. The direct flow, V_{direct}, is the volume of blood that entered during diastole and transits directly to the aortic outlet during the following systole, thus residing less than one heartbeat in the LV. The retained volume, $V_{retained}$, is the part that entered during diastole that is not ejected during the following systole. The delayed volume, $V_{delayed}$, was already present in the LV at the beginning of diastole and is then ejected during the following systole. Finally, the residual volume, $V_{residual}$, was present in the LV before the filling starts and is yet not ejected in the next systole. In a formula, the V_{ED} is divided as

$$V_{ED} = V_{direct} + V_{delayed} + V_{retained} + V_{residual}. \tag{12.6}$$

The first two terms on the right-hand side are the volumes that are ejected during systole, the stroke volume, $SV = V_{direct} + V_{delayed}$, and by difference with (12.6), using (12.1), the other two terms are those remaining in the LV at the end of the systolic ejection, $V_{ES} = V_{retained} + V_{residual}$. At the same time, when the valves are well healthy and ensure the absence of any backflow, the stroke volume also corresponds to the terms that enter during diastole $SV = V_{direct} + V_{retained}$, meaning that $V_{retained} = V_{delayed}$, up to measurement errors. These linear relationships among the four sub-volumes in Eq. (12.6) allow recovering three of them from the knowledge of a single one (typically V_{direct} or $V_{residual}$, which is the most representative of blood transit). For example, the measurement of V_{direct} provides information of the percentage of blood that transits across the LV without residing in more than 1 heartbeat; it is then immediate to obtain the $V_{residual} = V_{ES} - SV + V_{direct}$ that stagnates at least two beats inside the chamber, and so on with the others. It was shown that the direct flow component was reduced in dilated dysfunctional LVs with respect to normal hearts while the residual component increases, testifying a detrimental flow transit and higher risk of thrombus formation (Carlhäll and Bolger 2010).

The analysis of flow transit and residence time is relevant for recognizing the efficiency of blood motion and helps to stratify the risk of thrombus formation. This evaluation is commonly performed by releasing a large number N of virtual particles identified by their coordinates $X_i(t)$, $i = 1 \ldots N$, and letting them move with the local velocity

$$\frac{dX_i}{dt} = v(X_i, t), \tag{12.7}$$

where the velocity vector field $v(x, t)$ is made available by the imaging methods or by numerical solution. The particles can be initially distributed evenly in the LV cavity at end-diastole, each one representing a small volume of blood. They are then tracked forward in time during the LV contraction to evaluate those that exit during systole; they are also tracked backward in time during the previous diastole to obtain those that entered during the previous diastole. Such an approach corresponds to a Lagrangian representation where individual particles are followed in time. Another systematic approach to analyze flow transit can be based on the Eulerian equivalent of Eq. (12.7). It can be restated as a transport equation for the concentration of a passive scalar, which corresponds to individual blood particles "marked" at a certain instant during the cardiac cycle. It is common to integrate the transport equation with a diffusive term that mimics the diffusion of the fluid particles as it occurs to regular fluid elements. Call $C(x, t)$ the concentration of particles, the transport-diffusion equation is

$$\frac{\partial C}{\partial t} + \nabla \cdot (vC) = D\nabla^2 C; \tag{12.8}$$

with the diffusive coefficient D that can be placed equal to the kinematic viscosity of blood, or to zero to reproduce a pure transport with no diffusion. In incompressible flow, the second term in Equ. (12.8) can be rewritten in the standard convective form using the identity, which is valid when the velocity is divergence-free. Equation (12.8) can be solved with relative ease, numerically, once the velocity field $v(x, t)$ is known. This can be solved, for example, starting from end-systole with the condition that $C(x, 0) = 1$ everywhere in the LV volume. The concentration will decrease after every heartbeat of a percentage that depends on the quality of blood wash-out. The space-average value of concentration $\overline{C}(t)$

$$\overline{C}(t) = \frac{1}{V(t)} \int_{V(t)} C(x, t) dV$$

is a curve that is $\overline{C}(0) = 1$ initially and decreases after every heartbeat (typically exponentially) while the original blood washes out. It is immediate to see that the concentration after one heartbeat, at the following systole, corresponds to the ratio $\overline{C}(T) = V_{\text{residual}}/V_{\text{ES}}$. Therefore, the sub-volumes of Eq. (12.6) can be recovered by the first term of the wash-out curve. The time profile of the concentration curve provides a comprehensive information on the wash-out process. In dilated ventricles, the curve decays more slowly than in normal hearts and in presence of stagnation regions the tail of the curve is sustained for a long time because the region with blood stasis is more difficult to wash-out. This approach can also provide spatial maps of concentration to identify regions with reduced or accelerated wash-out.

An approach based on equations like (12.7) or (12.8) can be extended to more complex properties, like residence time or its combination with shear stress, to weigh the measure of stagnation with the potential degree of biological activation for developing thrombus. Clinical studies along this line are still at an early stage; however, they represent a promising technique for providing quantitative measures of the risk of thrombus formation and eventually modulate the anticoagulation therapy in subjects at risk (Seo et al. 2016).

(ii)   *Dynamic aspects: Hemodynamic forces and kinetic energy*

The dynamical actions exchanged between blood and tissue are represented by wall shear stress and pressure. Wall shear stresses are expected to be relevant when the behavior of blood motion is sensed from receptors on the tissue that trigger possible LV adaptations through complex processes of mechano-transduction that are still marginally understood (Pasipoularides 2015). Pressure has a more direct mechanical role on ventricular function, which was previously described by the pressure–volume diagram (Fig. 12.4) and Eq. (12.4). However, that description refers to the averaged value of pressure in the ventricular chamber and does not account for its spatial distribution. It should be reminded that heart function is that of creating blood flow. And we have seen multiple times in the previous chapters that fluid motion develops in virtue of the presence of a gradient of pressure, ∇p, whose value during time represents the dynamic coupling between myocardial activity and blood flow generation.

Pressure gradients drive blood motion during both ventricular ejection and ventricular filling as shown in Fig. 12.7. They represent the final result of LV deformation and play a central role in cardiac function that ultimately drives blood flow. Moreover,

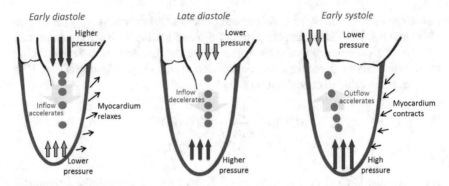

Fig. 12.7 Relationship between pressure gradient and flow acceleration in phases of the cardiac cycle

flow-mediated forces influence and participate in cardiac adaptation in presence of pathologies. Intraventricular pressure gradients are known from the literature (with measures made by catheters in animals) to have a fundamental influence on LV function (Courtois et al. 1988; Guerra et al. 2013). Despite their potential relevance, intraventricular pressure gradients have never been utilized in clinical cardiology due to the complexity of their acquisition that required invasive procedures.

The usage of pressure gradients has been recently renewed with the introduction of novel imaging techniques able to estimate the intraventricular blood velocities non-invasively (Arvidsson et al. 2018; Eriksson et al. 2017; Pedrizzetti et al. 2016). Indeed, once the intraventricular velocity field $v(x, t)$ is known, the ∇p field can be obtained after rearrangement of the Navier–Stokes equation as

$$\nabla p = -\rho \left(\frac{\partial v}{\partial t} + v \cdot \nabla v \right) + \mu \nabla^2 v. \tag{12.9}$$

In alternative to (12.9), the relative pressure field, up to a constant value (e.g. the average pressure) that can change in time, can be obtained by solving Poisson's equation

$$\nabla^2 p = -\rho \nabla \cdot (v \cdot \nabla v) = \frac{\partial v_i}{\partial x_k} \frac{\partial v_k}{\partial x_i}; \tag{12.10}$$

obtained by taking the divergence of the Navier–Stokes equation; last equality— that assumes summations over repeated indices- being valid in incompressible flow having $\nabla \cdot v = 0$. When solving (12.10), however, care must be taken in imposing appropriate boundary conditions because this is a second-order equation on pressure. Therefore, the average pressure gradient (tri-linear terms in pressure), which is often the property of main interest, is the solution of the homogeneous Laplace operator, $\nabla^2 p = 0$, and follows from the boundary conditions only.

Equations (12.9) and (12.10) are differential equations that are valid at every position inside the ventricular blood pool. However, clinical evaluations are principally interested in descriptors of the overall function and to global values, like the pressure gradient integrated over the LV volume. An integral dynamic property, known as hemodynamic force, is defined by the volume integral of momentum

$$F(t) = \rho \int_{V(t)} \left(\frac{\partial v}{\partial t} + v \cdot \nabla v \right) dV \cong \overline{\nabla p}(t) V(t), \tag{12.11}$$

which corresponds to the entire force exchanged between blood and surrounding tissue, including the viscous stresses. These are largely negligible and the hemodynamic force can be practically equated to the volume-average pressure gradient $\overline{\nabla p}(t)$ multiplied by the LV volume. At this stage, it is useful to recall the formulation (5.3) for the expression of momentum in integral form, which tells that the hemodynamic force vector (12.11) can be evaluated by an integral

$$F(t) = \rho \int_{S(t)} x \left(\frac{\partial v}{\partial t} \cdot n \right) + v(v \cdot n) dS; \tag{12.12}$$

on the surface S bounding the fluid volume. Therefore, the hemodynamic force can be obtained on the basis of the boundary dynamics without the need to know the velocity field inside the chamber. The bounding surface contains both the closed part, where fluid velocity coincides with that of the endocardium and the open (valvular) boundary where velocity is that effectively of blood. The approach based on (12.12) has the advantage of not requiring the measurement of blood velocity inside the volume and can be used with most standard imaging technologies that are dedicated to the visualization of tissues only. Interest in hemodynamic forces follows from the fact that this represents an indicator based on fluid dynamics; as such, it may be able to detect the presence of a sub-optimal cardiac function even before the tissues have developed measurable modifications.

In addition to the more established analysis of flow transit and dynamical interaction, there are further aspects of intraventricular flow that may be employed when describing the efficiency of LV function. Some authors proposed to evaluate the kinetic energy

$$\text{KE} = \frac{1}{2} \rho \int_{V} |v|^2 dV; \tag{12.13}$$

to describe the energetic level of LV flow; other authors suggested other properties (size, position, strength) related to the vortex formation process. On the energetic perspective, a possibly relevant property is the amount of dissipation of kinetic energy, given by the time integration of the rate of kinetic energy dissipation given in Eq. (6.2), that reflects the energetic efficiency of the blood flow pattern. The amount of energy dissipated by viscous friction inside the LV is certainly negligible with respect to that lost along the entire systemic circulation loop. Nevertheless, high levels of dissipation are commonly imputable to the lack of efficiency, sometimes including turbulence-induced fluctuations of pressure and wall shear stress, which may represent uncomfortable conditions for the cardiac function. These disturbing phenomena can trigger physiological feedback and eventually stimulate cardiac adaptation (Pedrizzetti et al. 2014) although no clinical evidence is available, yet.

12.4   Fluid Dynamics in Cardiac Pathology

Pathologies of the left ventricle can be roughly classified, by a mechanical viewpoint, in different classes. One important class of disease is that due to a weakened contraction imputable to *defects of myocardial perfusion* or ischemia. As discussed earlier in Sect. 11.2, this is a vascular pathology pertaining to the coronary arteries; however, it affects directly the LV function and will be discussed below in this respect. A wide class of cardiac pathologies can be described as general *mechanical dysfunction* associated with the inability to ensure a proper rhythm to cardiac function. A mechanical dysfunction can involve either systolic contraction or diastolic relaxation or both, and it can progressively lead to heart failure. The syndrome of heart failure is a general cardiac impairment that can develop as a primary disease or it can follow as a secondary effect after most other cardiac dysfunctions. A further class of pathologies is related to the presence of *electrical dysfunctions*. Some of these are purely neurological defects imputable to abnormalities of the electrical conduction system like arrhythmias and atrial fibrillation; their analysis is out of the scope of this book and they will be only mentioned in conjunction with other diseases. Other electrical dysfunctions give rise to improper contraction or relaxation, particularly to a lack of mechanical synchrony during the cardiac mechanical activity; they are broadly included in the general class of mechanical dysfunction in the discussion below as they often pave the road toward the heart failure. A further class of dysfunctions is that whose primary cause is imputable to *pathologies of cardiac valves*; there are discussed with more details in the next chapter. It should altogether be kept in mind that all such pathologies are interrelated and the classification reported above is driven by the present mechanistic viewpoint of cardiac function, especially of fluid dynamics, and may not reflect a classification associated with clinical scenarios.

(i)   *Defects of myocardial perfusion (ischemia)*

The most known pathology of the left ventricle is ischemia, whose extreme consequence is myocardial infarction: a regional loss of myocardial contraction that is a consequence of the reduction of myocardial perfusion from oxygenated blood due to coronary stenosis. This defect is a byproduct of vascular disease; because the reduction of myocardial perfusion follows from the reduced blood flow through a coronary artery due to its partial or total stenosis. The myocardial territory served by that vessel receives less oxygen allowance and becomes less able to perform contraction especially under high demand as during exercise.

The ischemic disease is commonly considered a systolic dysfunction because the myocardium is unable to properly contract during systole. However, by first principles, this is a vascular disease that should be treated at the vascular level as we have discussed earlier in Sect. 11.2. Nevertheless, its symptoms are noticed in terms of the inability of the myocardium to contract properly and it is commonly diagnosed by cardiac evaluation. Ischemic diseases present themselves with a reduction of the EF; this reduction is mostly due to regional contractile defect localized in the poorly perfused myocardial region, which can be recognized by cardiac imaging methods allowing visualization and quantification of myocardial motion. When this defect is small, it can be difficult to recognize under normal conditions and may become appreciable only under stress conditions, thus requiring cardiac evaluation performed under the exercise of pharmacologic stress. In the alternative, perfusion defects can be evaluated by perfusion imaging techniques, available in nuclear imaging, MRI, and, sometimes, echocardiography. When suspected, the presence and relevance of a coronary stenosis are eventually evaluated by coronary angiography as discussed in Sect. 11.2.

Intraventricular fluid dynamics is also affected by myocardial ischemia. Blood near a segment characterized by reduced motility is often more stagnant, especially when this is in the apical region. This implies a reduction of wash-out and increased risk of thrombi. It also alters the distribution of wall shear stress about the ischemic region, and it creates an imbalance in the intraventricular forces with the development of transversal pressure gradients that can give an excess of stress in some regions, that can be located even distant from the infarcted zone. Anomalous walls shear stresses and alteration of hemodynamic forces can progressively induce feedback and ventricular adaptation that may modify the LV geometry with potential further pathological implications.

Ischemia is typically solved by coronary endovascular surgery. However, when the solution is not complete, for example when there are multiple stenoses, some ischemia may remain and give ventricular imbalances. Similarly, when the ischemia has lasted for too long time, some regions of the myocardium may not be able to fully recover its contractile ability. The persistence of such imbalances may induce ventricular adaptation and mechanical dysfunction up to heart failure as discussed below.

(ii)   *General mechanical dysfunction (heart failure)*

Heart failure (HF) is the principal social threatening cardiac progressive dysfunction. It presents either as a primary pathology or as a consequence of numerous (almost all) primary diseases. It can be a consequence of partly recovered ischemia; it can follow electrical dysfunctions that do not allow a synchronous contraction; it can simply due to varied stiffness/thickness in the myocardium (for example due to hypertension or to fibrosis) that does not allow a uniform relaxation, to cite a few examples. On the other hand, it can develop as a primary disease following poor medical conditions. In any case, heart failure is the terminal stage of a progressive disease associated with impaired cardiac function.

The clinical syndrome of heart failure is associated with the development of ventricular remodeling: a modification of ventricular geometry that progressively alters its functional parameters whose final stage is the LV dilatation, known as dilated cardiomyopathy (DCM). Remodeling represents a physiologic adaptation feedback that often does not lead to a stable configuration rather to a progressively worsening of the cardiac function and eventually to failure. Despite modern treatments, hospitalization and death rate remains high; in the recent past, nearly 50% of people diagnosed with heart failure dying within 5 years (Lloyd-Jones et al. 2002).

The physiological causes leading to LV remodeling are mainly ascribed to an increase of stress on the myocardial fibers (around an ischemic area, or because of hypertension, etc.), which stimulates the growth and multiplication of cells giving rise to an increase in muscular thickness (hypertrophy) or extension (local dilatation). However, this picture is unable to differentiate patients exhibiting differences in LV structure and function, it is not consistently predictive of the future risk of cardiac remodeling and does not clarify how a regional disease rapidly remodels the LV as a whole. The availability of predictive models that can forecast the progression or reversal of LV remodeling following initiation of therapeutic interventions would be invaluable for overall risk stratification, improvement of preventive healthcare, and reduction of the perspective social burden.

The different stages of HF are depicted in Fig. 12.8. Heart failure is most commonly associated with ventricular dilatation (dilated cardiomyopathy, DCM). In this case, the myocardium is stretched and becomes thinner. The heart muscle contracts very little and is able to eject a sufficient SV with small contraction because of the large volume. The EF is significantly reduced, and we talk about HF with reduced ejection fraction (HFrEF), also referred to as systolic heart failure. Primary ventricular dilatation can develop as a result of volume overload (for example, due to regurgitation across the aortic valve during diastole, as discuses in the next chapter)

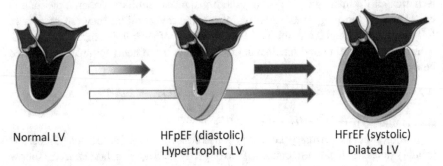

Normal LV   HFpEF (diastolic)   HFrEF (systolic)
                     Hypertrophic LV   Dilated LV

Fig. 12.8 Types of remodeling and heart failure (credit: adapted from Messerli et al. JACC Heart Fail. 2017;5:543, with permission from Elsevier)

causing an eccentric effort in the myocardium, or it can be due to a general weakening of the myocardial tissue. In HFrEF, the intraventricular fluid dynamics is very weak; the SV is a small percentage of the chamber volume. Typically, blood flow takes either a continuous weak rotary motion, when the inflow is aligned to feed the central vortex, or it presents a weak turbulence. In both cases, flow transit is featured by stasis and high thrombotic risk. Intraventricular hemodynamic forces are reduced in entity and present an incoherent pattern with varying directions.

Another type of HF is associated with thickening and/or stiffening of the myocardium. The ventricular volume is about normal and the pumping parameters are also normal but the ventricle does not relax properly during ventricular filling because of its stiffness. The EF is thus preserved, usually because the ventricle is hypertrophic (hypertrophic cardiomyopathy, HCM) and the inward thickening helps to support systolic ejection. This pathology, which is more difficult to recognize, is sometimes called HF with preserved ejection fraction (HFpEF), also referred to as diastolic heart failure. Either myocardial hypertrophy or stiffening can develop as a result of pressure overload (for example due to hypertension, or to a stenosis in the aortic valve) causing a concentric effort in the myocardium, or it can occur for multiple causes including fibrosis in the myocardium that can be associated to genetic predisposition. Sometimes hypertrophy/stiffening represents a preliminary dysfunctional stage that evolves to HFrEF at its later stage. Intraventricular blood flow in HFpEF subjects is similar to normal; however, dynamical differences reflecting the altered flow pattern are expected including reduction of the entity of hemodynamic forces; although, definitive results are not available, yet.

Despite the general considerations reported above, the causes leading to LV remodeling are still largely incomplete. During the progression, there are changes in the pumping function. These can be noticed by a reduction of systolic trust, as well as by changes in the relative intensity between diastolic E-wave and A-wave, with an extra-burst by atrial contraction when the early filling is insufficient, or alteration of the timing of acceleration and decays of individual phases. Clinicians use the combination of numerous indicators trying to figure out the specific pathological scenario; however, a comprehensive mechanical picture is still missing.

It has been recently shown that alterations in the intraventricular fluid dynamics are observable well before the tissue has undergone noticeable often irreversible changes. Given the incompressible nature of blood, in a cardiac chamber that is filled with blood, every segment is somehow in touch with the others through the column of fluid between them, as a result, the blood inertia associated with the rapid acceleration-deceleration about one region can instantaneously influence distant regions. The role of flow on cardiac remodeling has been considered in the past only through global indicators like volumetric changes, the inflow velocity of E- and A-wave, or combinations thereof. The absence of more specific fluid dynamics indicators is mainly due to the lack of technologies able to evaluate intraventricular fluid dynamics with sufficient ease and reliability.

Blood flow responds immediately to any small change in the surrounding conditions. Therefore, for its nature, it is expected to be one of the first dynamical phenomena that display alterations during the initial stage of dysfunction, whereas

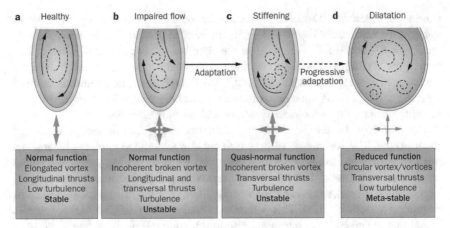

Fig. 12.9 Flow-mediated path toward heart failure (credit: Pedrizzetti et al. Nature Reviews Cardiology 2014. https://doi.org/10.1038/nrcardio.2014.75)

changes in the tissue require some time to develop at a sufficiently extended degree to be noticeable. Normal intraventricular fluid dynamics is known to be associated with a physiologically stable cardiac function that does not lead to remodeling. Vice versa, a progressive disease corresponds to a physiologically unstable state that is expected to depart and lead away from normality. As shown schematically in Fig. 12.9, an alteration of intraventricular fluid dynamics induces alteration of forces and shear stress on the tissue, these can trigger adaptation feedbacks and bring to progressive dysfunction (Pasipoularides 2015; Pedrizzetti et al. 2014). In an initial phase, the alteration of flow-mediated stresses may lead to stiffening of the myocardial tissue that is sometimes associated with the increase of myocardial thickness (hypertrophy). This can be a condition going to HFpEF, or it can be just a passage toward progressive tissue dilatation with a further reduction of LV function and eventually going to the more common HFrEF.

Therapies for heart failure are complicated as they should go to the cause leading to remodeling. Moreover, HF often involves dysfunction in physiologically related organs and, therefore, precise guidelines and therapies are varied.

Heart failure can also follow in consequence of the presence of a mechanical dyssynchrony in the myocardial activation. It can be a lack of proper timing between the cardiac chambers, or can be a dyssynchrony between the different regions inside the ventricle. The causes leading to dyssynchrony can originate directly from alteration in the electrical conduction system, or they can follow for different levels of impairment in the tissue that varies its response in entity and timing. It is not uncommon that dilated ventricles present a mechanical dyssynchrony. Multipoint pace makers are able to improve the LV function by restoring its synchrony during contraction and relaxation. They were shown to be one successful option in many cases, especially when HF is associated with a disturbed electrical activity (either as a cause or a consequence of HF). This approach, called cardiac resynchronization therapy (CRT), requires the definition of stimulation intervals in the pace-maker to ensure the optimal therapeutic outcomes. Typically, they can be chosen by electric conduction optimization or through synchronization of myocardial tissue motion. However, this approach is prone to substantial improvements that include on one side the identification of optimal stimulation points, on the other the definition of optimal stimulation timings for the different places. Fluid dynamics offers a global perspective to define the proper contraction pattern, by ensuring that the hemodynamic forces are maximized and properly aligned along the base-apex direction. However, studies are currently in progress to verify effective clinical relationships (Arvidsson et al. 2018; Eriksson et al. 2017; Pedrizzetti et al. 2016).

This concept can, however, be generalized to evaluate the normality of cardiac function after the acute cause that may, or may not, lead to heart failure. These include endovascular prosthesis, valvular repair or transplant, and so on. Intraventricular fluid dynamics appears as the first mechanical factor modified after the alteration of cardiac function, even when they are minor and unnoticeable. As such, it appears a promising central element for the prediction of progressive disease or of therapeutic outcomes.

Chapter 13
Cardiac Mechanics II: Heart Valves

Abstract Cardiac valves represent fundamental elements of cardiac function which directly interact with blood flow. The fluid dynamics of the aortic valve and of the mitral valve are here described in the normal operating conditions, first. Flow features in presence of valvular diseases are then discussed in conjunction with the technology available to evaluate them in the clinical setting. Valves are subjected to two principal pathologies: stenosis is a narrowing of the valve that makes the flow through it more difficult; insufficiency is a loosening of the valvular tissue that does not guarantee complete closure and allow backflow or regurgitation. The therapeutic solutions were principally those of a surgical approach; since recently, transcatheter valve replacement is becoming the main option. In both cases, therapy has an impact on blood flow patterns that can influence the long-term outcome. The chapter terminates with a brief description of the main congenital cardiac diseases, the tetralogy of Fallot and hypoplastic left heart syndrome, and the associated modifications in the blood flowing through.

13.1 Cardiac Valves

The heart contains four valves, as sketched in Fig. 13.1. Two of them are atrioventricular valves, the mitral valve on the left side and the tricuspid on the right side; the other two valves are for the communication from the ventricle to the circulation, the aortic valve and the pulmonary valve for the left ventricle (LV) and right ventricle (RV), respectively. The main function of the cardiac valves is to allow flow in one direction and prevent backflow.

During systole, the ventricles contract and eject blood through the aortic and pulmonary valves, for the LV and RV, respectively, while the atrioventricular valves are closed. Ventricular contraction is made of an inward motion of the ventricular endocardial surface, combined with a shortening of the base-apex length that, given that the apex is relatively fixed, is obtained by the motion of the entire valvular plane downward. Vice versa, during diastolic ventricular expansion, the ventricles expand and the valvular plane moves upwards. This upward-downward motion creates a relative velocity at the valve that supports ventricular filling-emptying and helps

© The Author(s), under exclusive license to Springer Nature Switzerland AG 2022   211
G. Pedrizzetti, *Fluid Mechanics for Cardiovascular Engineering*,
https://doi.org/10.1007/978-3-030-85943-5_13

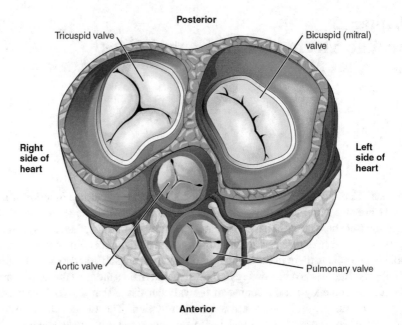

Fig. 13.1 Valvular plane containing the 4 cardiac valves, as seen from the top of ventricles. (credit: OpenStax College, CC BY 3.0, via Wikimedia Commons)

anticipating valvular opening and closure. It must otherwise be reminded that the blood velocity measured just above and below the valves can be non-zero even when the valves are closed because the motion of the valvular plane transports blood with it.

Despite their overall common function, cardiac valves present important differences, due to the actual anatomical position and to the fluid dynamics operating conditions. The therapeutic solutions can also be very different. We discuss here the two main valves sited on the left heart, as the valves on the right side are much less studied and their consideration are still mostly borrowed from the left ones.

13.2   Aortic Valve

The aortic valve is situated between the end of the left ventricular outflow tract and the origin of the aorta. It consists of three leaflets of approximately triangular shape sometimes referred to as cusps. One side of each triangular leaflet is attached to the fibrous annulus embedded inside the muscular walls of the LV and represents the hinge for the leaflet motion. The aortic valve is crossed by flow ejected from the LV into aorta during systole and it remains closed during diastole when its function is that of preventing backward flow into the LV. Once it is closed, the cusps present a coaptation in their tip portions that remain aligned downstream and resist backward displacement.

The anatomy and function of the aortic valve have inspired studies for the past 600 years beginning with Leonardo da Vinci who studied the aortic valve and the role of the sinuses of Valsalva. The Valsalva sinuses are dilatations in the aortic wall, just behind the valve, in correspondence with each of the semilunar cusps. Generally, there are three aortic sinuses, the left, the right, and the posterior, each one in correspondence of a leaflet. The left aortic sinus gives rise to the left coronary artery, and the right aortic sinus gives rise to the right coronary artery, while no vessel originates from the posterior aortic sinus, which is known as the non-coronary sinus. A sketch of the geometry of the aortic valve and the aortic root is shown in Fig. 13.2.

Blood flows across the valve are influenced by its geometric properties. Given the enlargement at the Valsalva sinuses, and the close-to-triangular shape of the open valve orifice, the systolic flow gives rise to a boundary layer separation from the nearly straight edge given by each of the open leaflet and detaches downstream as a free shear layer that rolls-up forming a vortex structure that develops recirculatory motion in the Valsalva sinuses (see Fig. 13.2, right panel). The role of vortex formation in the sinuses is not completely understood, yet. This backflow was initially considered to help the leaflet closure at the end of systole; it is also expected to facilitate the flow into the coronaries. More likely, the coronary flow is principally driven by the backflow that develops near the boundary during diastolic deceleration, when the bulk flow, at the center of the aorta still moves downstream, and by the reflected

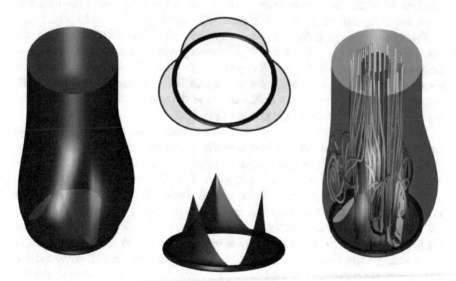

Fig. 13.2 Sketch of the aortic valve and aortic root made from a mathematical model. Three-dimensional view (left picture), isolated valve in open position (lower picture in the central panel) and cross-section of at the level of the Valsalva sinuses (upper picture). Streamlines at peak systole shows the recirculation in the sinuses

Fig. 13.3 Turbulent flow
across a bi-leaflet prosthetic
valve reconstructed from
high-resolution numerical
simulations (credit: Prof.
Marco Donato de Tullio,
Politecnico di Bari, Italy,
own scientific work)

pressure wave. Surely, the presence of the Valsalva sinuses prevents the leaflet to touch the aortic wall and to close of the coronary entrance.

The aortic jet presents itself as a turbulent jet with a Reynolds number that can reach about 10,000 in normal condition (velocity near 2 m/s crossing an orifice with mean diameter about 2 cm). It is probably the only fully turbulent flow in the circulatory system. The Strouhal number is about 10^{-2}, thus the jet is well above 10 diameters long. The complexity of blood flow at the aortic root is perceived by the high-resolution numerical results shown in Fig. 13.3.

The aortic jet exists from an orifice into the center of the aorta thus, despite its strength, in normal conditions the high-speed blood does not impact directly onto the aortic walls. The jet exists straight in the aortic root and develops helical streamlines when going through the aortic arch as shown in Fig. 13.4 (left side).

Normal aortic valve is tricuspid; however, a significant percentage of the population (about 2%) is born with a bicuspid aortic valve (BAV) where two leaflets are not fully separated or they are totally fused as one. One possible effect of BAV is the reduced orifice size when the valve is open, giving rise to an even stronger jet and possibly higher resistance to the ejection requiring an extra effort to the LV with consequences similar to what happens in valvular stenosis (discussed below). Another important possible phenomenon related to a BAV is the asymmetric opening of the unequal leaflets, which may deviate the jet toward the aortic wall. This increases the risk of damaging the wall that may weaken and facilitate the development of aneurisms in the aortic root. An example of flow recorded in presence of a BAV that

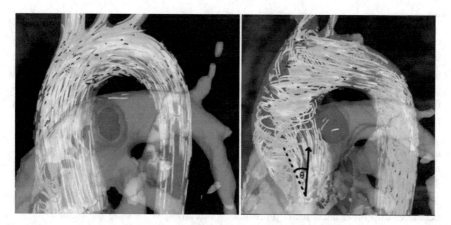

Fig. 13.4 Aortic jet recorded from MRI, jet with a normal alignment (left) and deviated jet in the presence of a bi-leaflet aortic valve (right). (credit: Bissell, et al. Circ. Cardiovasc. Imaging 2013;6:499, with permission from Wolters Kluwer Health Inc.)

deviated the direction of the jet is shown in Fig. 13.4 (right side). BAV subjects can have a normal life; however, given the additional risk factors, they must be monitored to ensure the absence of progressive diseases development.

13.3 Pathologies of the Aortic Valve

Major valvular pathologies can be roughly grouped, from a mechanical standpoint, as those due to valvular *stenosis* or to valvular *insufficiency*.

Valvular stenosis is a reduction of the valvular orifice due to calcification of the valve leaflets that makes them less elastic and more difficult to open. Valvular stenosis thus reduces the effective orifice area and provokes a stronger jet at the entrance of the aorta, with velocities that can reach several meters per second, which means higher turbulence and risk of damage to the arterial wall when such jet has deviated.

The major consequence imputable to valvular stenosis is the higher energetic resistance to ejection: a higher pressure drop across the aortic valve that, at peak systole, is proportional to the square of velocity, see Eq. (6.14). This additional pressure loss can be significant (if velocity is in m/s, pressure loss in mmHg is given by $4v^2$) and represents an extra effort totally in charge of the LV as it occurs immediately at its exit. This means that the ventricle requires to build up a higher pressure (pressure overload) to get the same output in the aorta, and the LV myocardium is required to develop more force. Such a condition can likely give rise to LV hypertrophy, tissue stiffening setting up a possible path toward heart failure as discussed in the previous chapter.

The other major pathology of the aortic valve is insufficiency. In valvular insufficiency, the leaflets are looser or the valve is dilated, and leaflets coaptation is insufficient; as a result, the leaflets are unable to properly close the valve during diastole giving rise to valvular *regurgitation*. This means that, during LV filling, when the LV pressure decreases and blood flow in through the mitral valve, some flow also enters into the LV back from aorta. Therefore, part of the net LV pumping effort is wasted because a percentage of the ejected blood returns into the LV itself.

Measuring the entity of the regurgitated volume is the principal mean to assess the severity of aortic insufficiency. It can be performed by phase-contrast MRI, recording the velocity during diastole across a plane just above or below the valve; this is the most accurate option, although relatively time-consuming procedure, which is performed on patients requiring an accurate evaluation. A simpler, less accurate, approach used for a preliminary screening is feasible by color Doppler echocardiography that permits looking at the color map of the vertical component of blood velocity. The image is commonly recorded at peak diastole, the regurgitating jet downstream the valve is not measurable because velocities are too high and disturbed; instead the color Doppler image "proximally" to the regurgitating orifice typically presents a regular pattern corresponding to a smoothly converging flow and it can somehow be analyzed. The most common method, called proximal iso-surface velocity area (PISA), hypothesizes that—far enough upstream the orifice—the flow converges across a series of concentric half-spheres; therefore the value of the Doppler (vertical) velocity, v_{Doppler}, at an upstream distance R on the axis can be assumed to be equal to the radial velocity over a hemispherical shell as shown in Fig. 13.5. In such a pattern, the regurgitating discharge is obtained, by continuity, as that crossing the shell.

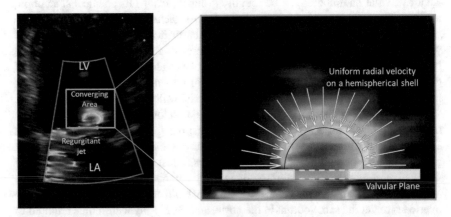

Fig. 13.5 PISA method to estimate the regurgitating flow from color Doppler echocardiographic image proximal to the insufficient orifice. This procedure, shown here for mitral regurgitation, applies equally to the aortic valve where the proximal area is inside the aorta and the jet into the left ventricle

$$Q_{\text{peak}} = 2\pi R^2 v_{\text{Doppler}};\qquad(13.1)$$

This gives the regurgitating flow rate at peak diastole. The regurgitated volume, V_{regurg}, is estimated by assuming a proportionality between regurgitant flow across the aortic valve and regular inflow across the mitral valve. The time profile of the latter can be recorded (by pulsed-wave Doppler) its peak value (v_{peak}) and velocity time integral (VTI) are then evaluated, an operation that can be performed in most echographs. The proportionality then allows to estimate $V_{\text{regurg}} = \text{VTI} \times Q_{\text{peak}}/v_{\text{peak}}$. The entire PISA approach is very approximate, some further improvement has been introduced by most vendors of ultrasound equipment by using 3D color Doppler data and corrections for irregular orifices. It has the merit to be a quick procedure feasible routinely; nevertheless, it should be repeated to improve the reliability of results and used as a preliminary information only and not as a rigorous measurement.

The entity of regurgitation is not the only matter associated with the clinical severity of the insufficiency. Aortic regurgitating jet can conflict with the mitral inflow as shown in Fig. 13.6, giving rise to turbulence and disturbed LV filling that may further affect the LV function. Some studies suggested that the severity of aortic regurgitations could be measured by evaluating the degree of irregularity of the intraventricular flow during diastole; however, this was only a suggestion and clinical results are still inconclusive (Pedrizzetti and Sengupta 2015).

Because of aortic insufficiency, the LV tends to dilate for the extra volumetric load coming from the regurgitated blood (volume overload). At the same time, the

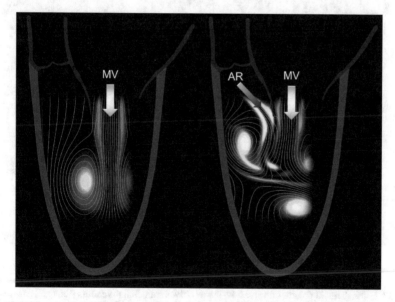

Fig. 13.6 Flow in the LV in presence of aortic valve regurgitation (credit: adapted from Pedrizzetti and Sengupta. Eur Heart J, Cardiovasc Imag 2015;16:719, with permission from Oxford University Press).

reduction of net flow downstream in the aorta induces metabolic feedback to stimulate the LV pumping to allow the necessary blood in the circulation. This requirement of an abnormal extra effort, demanded to an LV that was already increasing its volume, sets again the path toward heart failure.

The therapeutic solutions to aortic stenosis as well as aortic regurgitation are those of surgical valvular repair or, most commonly, valvular replacement. Surgical aortic valve replacement contemplates the substitution of the diseased valve with a prosthetic one that is directly sutured in its place. This type of surgery may also include the substitution of the aortic root with a prosthetic vessel.

Several types of prosthetic mechanical valves were introduced in the past and are still designed; a few examples are shown in Fig. 13.7. Currently the most common is the bi-leaflet mechanical valve, which ensures life-long duration, although the leaflet geometry differs substantially from those in a natural valve and may alter the flow pattern downstream. Further designs are still under development with the objective of mimicking the natural geometry. However, due to the hardness of the material, mechanical valves produce a phenomenon known as hemolysis: red blood cells are subjected to rupture during the interaction with hard mechanical elements inducing

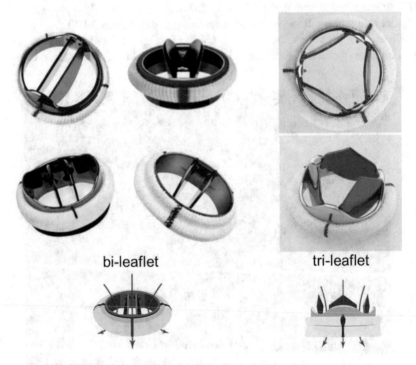

Fig. 13.7 Mechanical prosthetic valves: bi-leaflet valves from major vendors (left) and a tri-leaflet prototype, with a sketch of fluid direction (bottom) (credit: adapted from Kheradvar et al. Ann Biomed Eng 2015;43:844 and from Robinson et al. Animal Models for Cardiac Research 2015 https://doi.org/10.1007/978-3-319-19464-6_27, with permission from Springer Nature)

coagulation phenomena and a higher risk of thrombus formation. For this reason, they also require life-long anticoagulant medication. Mechanical valves also largely alter the fluid dynamics downstream of the valve. The aortic jet presents multiple shear layers with an altered vortex formation process and higher turbulence. The relationship between the three leaflets and three Valsalva sinuses is broken thus other contraindications may accompany such an implant. Occasionally, the mechanical valve can also give rise to the formation of gas bubbles and cavitation. In the circulatory system, cavitation is not frequent, but it has been sometimes observed in mechanical heart valves. This phenomenon presents when pressure is low and the sharp local changes due to the mechanical element can create small regions where the absolute pressure approaches zero (reaches the tension vapor in the liquid). In this case, low-pressure bubbles of gas form and they can implode as soon as they move to place with higher pressure and likely damage the surrounding tissues; bubbles can also coalesce and produce emboli. Bubble formation downstream valves is usually monitored by imaging methods and the risk of cavitation must be adequately considered in the design of mechanical prosthesis (Qian et al. 2019).

A more natural alternative is that of biological valves that do not require anticoagulant and better mimic the original natural geometry to reproduce a more natural fluid dynamics behind the valve. Biological valves, shown in Fig. 13.8, on the other side, are not guaranteed for a life-long duration although technological improvements give confidence for their reliability.

With the advent of the trans-catheter approach to valvular replacement, which rapidly grew from early 2000s (Bourantas and Serruys 2014), the surgical procedure of valvular replacement has become less frequent. Since then, the most widely used

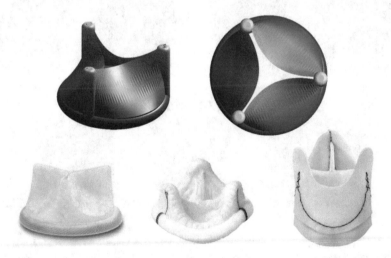

Fig. 13.8 Bio-prosthetic valves: schematic design (top) commercial valves with different scaffolds and prosthetic annulus (bottom). (credit, top images: created by sjpiper145 licensed under CC Public Domain Dedication license; bottom images: adapted from Piazza et al. JACC Cardiovascular Intervention 2011;4:721, with permission from Elsevier)

solution that avoids open surgery is the trans-catheter aortic valve implant (TAVI, or equivalently called trans-catheter aortic valve replacement TAVR). In TAVI, the valve is placed inside an endovascular prosthesis that can be positioned by catheter avoiding surgery. In this procedure, the natural diseased valve is initially squashed at the wall, and then the new valve is expanded and placed over the previous one, as shown in Fig. 13.9 (top sequence). Initially, TAVI was used for a patient at risk of open-chest surgery only. Later on, advantages have been so numerous that it is recommended in most cases. The number of TAVI prostheses available is continuously increasing, a few examples are shown in Fig. 13.9 (bottom panel).

The resulting fluid dynamics after TAVI is very similar to that of a biological valve and does no exhibit drastic changes from that of a natural aortic valve. A critical effect can be the presence of para-valvular blood leakage when the new valve does not adhere perfectly to the side tissues and allows some blood to pass in the small gaps between the tissue and the implanted valve. This may give rise to a para-valvular regurgitation. After years of experience with usage of TAVI, it is occasionally required to intervene over a previously installed failing trans-catheter

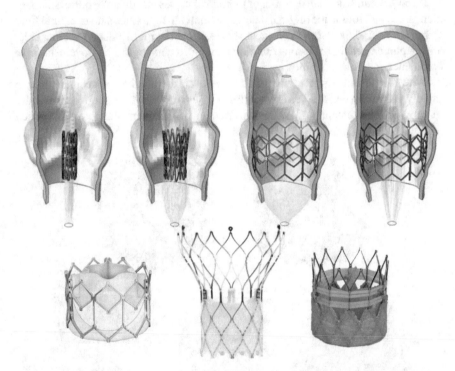

Fig. 13.9 Trans-catheter aortic valve implant procedure (top panel), a few examples showing different typologies (bottom panel). (credit, top images: Auricchio et al. Computer Methods in Biomechanics and Biomedical Engineering 2014;17:1347, with permission from Taylor & Francis; bottom left image: image courtesy of Edwards Lifesciences Corporation; bottom center image: from Kheradvar et al. Ann Biomed Eng 2015;43:844, with permission from Springer Nature; bottom right image: courtesy of Prof. Arash Kheradvar, University of California Irvine)

valve by a novel TAVI to be placed on top of the previous one. This procedure, called valve-in-valve, presents similar criticalities in fluid dynamics terms, however, experience is still at an early stage.

13.4 Mitral Valve

Mitral valve is the bi-leaflet valve that connects the left atrial chamber to the left ventricle. The valve consists of two leaflets of unequal size, with a coaptation between the two that takes a D-shape, as artistically shown in Fig. 13.10. The anterior leaflet is the largest, positioned on the anteroseptal side, the right side of the mitral orifice next to the left ventricular outflow tract, while the smaller posterior leaflet is placed on the left side, close to the posterior-lateral wall.

The leaflets' edges are connected to the papillary muscles via cord-like tendons, called chordae tendineae, that prevent valvular opening toward the atrium. The chordae tendineae are required to hold the leaflets during systole in presence of a high-pressure difference between the LV, which develops a high systolic pressure, and the low left atrial pressure. While the aortic valve is inside a tubular shape vessel, the mitral valve is contained in the atrioventricular plane; here, the mitral valve is surrounded by a fibrous annulus, which approximates a hyperbolic paraboloid similar to a riding saddle, which modulates its shape during the heartbeat.

The transmitral flow is characterized by two impulses, the early filling wave (E) and the atrial contraction (A). Before the early filling, at the end of systole, the ejected flow decelerates and LV pressure is lower at the apex than at the LV base.

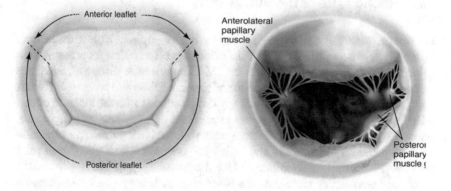

Fig. 13.10 Mitral valve, closed (left) and open (right) showing the chordae tendineae attaching the leaflets to the papillary muscles inside the ventricle (credit: curtesy of Dr. David H. Adams, Department of Cardiovascular Surgery, the Icahn School of Medicine at Mount Sinai, New York; illustrations with permission from Carpentier A, Adams DH, Filsoufi F. "Carpentier's Reconstructive Valve Surgery - From Valve Analysis to Valve Reconstruction", 2010, SAUNDERS, Elsevier)

At the passage between systole and diastole, the myocardial contractile elements are deactivated and release the stored elastic energy, this results in further pressure drop inside the LV whose response is the opening of the mitral valve. Both these two mechanisms about the transition between systole and diastole are associated with a lower pressure at the apex than at the LV base and both actively contribute to early ventricular filling impulse. Afterward, before the end of diastole, the electric stimulation starts with the atrial systole and the A-wave completes the LV filling. The relative entity of E and A-waves is an indicator of LV function. The E and A peaks of mitral velocity are usually assessed by Doppler echocardiography of the mitral inflow; typically, E-wave velocity is some greater than the A-wave; this ratio is reversed when the early filling is insufficient and additional effort is given by atrial contraction, suggesting diastolic dysfunction. This ratio is also reversed in normal fetal hearts before cardiac maturation.

The anatomic asymmetry of the mitral valve has a fundamental influence on the development of LV fluid dynamics. The vortex formation process is made of a distorted vortex ring that is stronger on the anterior side and weaker on the posterior; this arrangement induces the ring to deviate toward the posterior side (because the anterior side has a higher self-induced velocity, while the posterior side is slowed down by the image vorticity at the wall). As a result, the larger leaflet on the anterior side helps to redirect the blood flow along the lateral-posterior wall. The anterior vortex eventually occupies most of the LV cavity and ensures the development of a proper circulation inside the LV. Normal transmitral flow is usually laminar and relatively low in velocity (usually less than 1 m/s); nevertheless, the vortex formation creates vortical structures that are complex although not strictly turbulent.

13.5 Pathologies of the Mitral Valve

Like for the aortic valve, mitral stenosis, which is normally imputable to leaflets calcification, gives rise to a reduction of the orifice size. The mitral jet presents higher velocities and can be deviated inside the LV. This can create disturbed, even turbulent flow with higher energy dissipation and abnormal shear and pressure increase on regions of the wall. The narrower valve is also associated with the increase of transmitral pressure drop, with consequent impairment of LV filling, and higher atrial pressure. The increased atrial pressure can influence back pulmonary circulation, produce pulmonary congestion and higher RV pressure. These effects can set the path towards diastolic heart failure and RV dilatation.

Mitral insufficiency represents the other main pathology that occurs when the valve is unable to prevent backflow during LV contraction giving rise to mitral regurgitation. It can appear as a secondary effect to LV dilatation; in this case, the entire LV increases its volume and the mitral annulus also enlarges such that the leaflets are unable to cover the entire mitral area and close the orifice. Mitral insufficiency, however, frequently develops as a primary valvular disease in presence of mitral valve prolapse. Mitral prolapse is due to the growth of one leaflet that becomes

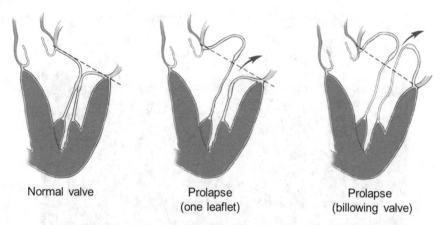

Normal valve   Prolapse   Prolapse
                              (one leaflet)   (billowing valve)

Fig. 13.11 Mitral valve prolapse and regurgitation (credit: curtesy of Dr. David H. Adams, Department of Cardiovascular Surgery, the Icahn School of Medicine at Mount Sinai, New York; illustrations with permission from Carpentier A, Adams DH, Filsoufi F. "Carpentier's Reconstructive Valve Surgery - From Valve Analysis to Valve Reconstruction", 2010, SAUNDERS, Elsevier)

wider, longer, and looser. The leaflet of the mitral valve bulges back into the left atrium pushed by the high LV pressure during systolic contraction, sometimes this phenomenon occurs to both leaflets (Barlow disease) that enter into the atrium like a parachute held by the chordae tendineae at the edges. Finally, mitral regurgitation can also be a consequence of chordae elongation and sometimes to their rupture that fails in the work of withholding. Prolapse is a frequent phenomenon giving no specific symptoms at its early stage. However, it must be monitored because, as shown in Fig. 13.11, eventually the loose leaflets may not properly close the valve and allow blood to flow backward into the left atrium producing mitral valve regurgitation.

The severity of mitral regurgitation can be evaluated by measuring the regurgitated volume with the same imaging methods (MRI or echography) previously described for aortic valve regurgitation.

Mitral regurgitation reduces the effectiveness of LV pumping because part of the stroke volume is not ejected into the aorta and flows backward into the left atrium. This induces metabolic feedbacks to increase LV pumping and stressing the LV, especially under exercise or stress. The most evident pathologic consequence of severe regurgitation is the dilatation of the left atrium, which must comply with the additional blood volume and is subjected to systolic LV pressure. When the atrial dilatation becomes important, mitral prolapse requires treatment.

Pharmacologic treatments to mitral valve diseases can rarely heal the defect. A surgical solution to mitral valve stenosis or, sometimes, prolapse is the replacement of the diseased valve with a prosthetic valve. As discussed for the aortic valve, the prosthesis can be either biological or mechanical. A prosthetic valve can significantly alter the intraventricular fluid dynamics with different flow patterns depending on the type, orientation, and position of the valve (Faludi et al. 2010). It was shown, see Fig. 13.12, that the symmetry of a mechanical bi-leaflet, in contrast with the

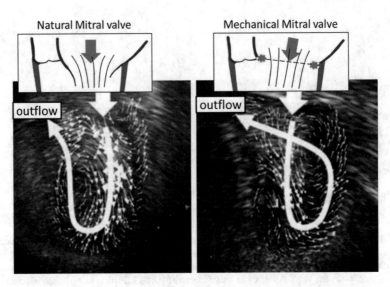

Fig. 13.12 Flow redirection with bi-leaflet mechanical valve in mitral position (credit: Pedrizzetti et al. Ann Biomed Eng 2010;38:769, with permission from Springer Nature)

natural asymmetry of the mitral valve, increases turbulence and may even reverse the vortical circulation inside the LV. However, as these observations are difficult to perform clinically, there are no indications of the consequences of such LV flow alterations.

The most common surgical option is mitral valve repair (MVR), which has evolved over the years and can be performed under different procedures (Carpentier et al. 2010). MVR is the primary choice for prolapse, although surgery is performed in the presence of stenosis as well. As shown in Fig. 13.13, MVR aims to recreate the natural valvular geometry removing the exceeding tissue and suturing the original tissue into a proper geometry. Often, MVR is performed including a new prosthetic mitral ring, replacing the older one that can be dilated, to enforce the appropriate dimension of the reconstructed valve.

The fluid dynamics after MVR can vary a lot depending on the details of the surgical procedure. The primary endpoint is represented by the reduction of regurgitation and the influence of repair to the intraventricular flow is rarely monitored. Nevertheless, the long terms outcome can vary substantially after similar MVR procedures and it is sometimes suggested that it is likely influenced by the flow pattern that develops after repair. However, systematic evidence in this sense are not yet available.

Trans-catheter mitral valve repairs (TMVR) are less common than they are for the aortic valve, because they present the complexity to anchor the prosthesis in the mitral plane, without a surrounding vessel as was available for the aortic valve to place the stent. This solution started to be available in the late 2010s, clinical experience is lower than for TAVI and it has to face additional technical challenges

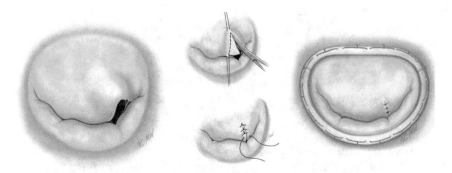

Fig. 13.13 Mitral valve repair with triangular resection of a prolapsed anterior leaflet and annuloplasty (credit: curtesy of Dr. David H. Adams, Department of Cardiovascular Surgery, the Icahn School of Medicine at Mount Sinai, New York; illustrations with permission from Carpentier A, Adams DH, Filsoufi F. "Carpentier's Reconstructive Valve Surgery - From Valve Analysis to Valve Reconstruction", 2010, SAUNDERS, Elsevier)

(Regueiro et al. 2017). Nevertheless, the applications of TMVR are rapidly growing and novel technological solutions are in continuous progress.

One endovascular solution for reducing regurgitation in mitral valve prolapse has been recently introduced. It consists of a "clip" (similar to a paper clip) inserted trans-catheter that sticks together the two leaflets thus transforming the wide prolapsed orifice into two small orifices, as shown in Fig. 13.14, which do not allow regurgitation when closed. This method is a trans-catheter version of a previous surgical solution called edge-to-edge repair. After the mitral clip insertion, regurgitations are normally reduced or eliminated; however, this treatment dramatically alters the intra-ventricular fluid dynamics, as demonstrated since the introduction of edge-to-edge repair (Redaelli et al. 2001). As shown in Fig. 13.14, the mitral jet transforms into two distinct jets diverging from the valve and impacting on the opposite walls, developing higher turbulence, varied shear stress, and intraventricular pressure gradients. The long-term clinical consequences of this alteration are still not verified. This solution is advised for critical mitral regurgitation conditions and for patients that cannot undergo other treatment options.

13.6   A Mention to Congenital Cardiac Disease

A number of diseases are imputable to congenital malformations of the heart, most of them related to pathological alterations of cardiac valves since the early phases of heart development. The topic of congenital heart diseases is wide and complex, and it is out of the scope of this basic text. However, for the sake of completeness, it is worth mentioning two exemplary situations among the most common (frequency about of 1 every 2000 children) severe congenital heart defects that are found in newborn children: tetralogy of fallot, and hypoplastic left heart syndrome.

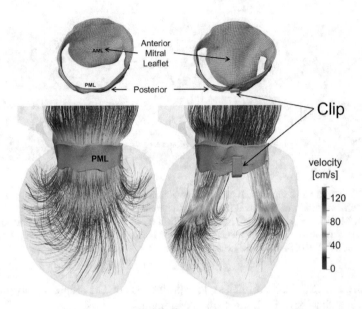

Fig. 13.14 Fluid dynamics before (left) and after (right) mitral valve edge-to-edge repair with mitral clip trans-catheter repair (credit: Caballero et al. Frontiers in Physiology 2020;11:432, CC BY)

The tetralogy of fallot (TOF) is a combination of four defects, interrelated and concurring, that directly influence the blood circulation in the heart. Each defect one can present with different degrees of severity and in different combinations. The common result is low blood oxygenation, which can give rise to cyanosis; for this reason, this defect is also called the "blue baby syndrome".

TOF is characterized by the followings malformations as graphically described in Fig. 13.15.

1. A defect in the interventricular septum that is not complete and allows passage of blood between RV and LV; this means that part of the non-oxygenated RV blood can enter the LV and be delivered into the primary circulation.
2. The pulmonary valve, at the RV outlet, is narrower thus reducing the amount of blood delivered toward the pulmonary circulation for oxygenation.
3. The aorta is displaced toward the right side because the basal part of the interventricular septum is absent; therefore, aorta can receive either the oxygenated blood from the LV and part of the non-oxygenated blood ejected by the RV.
4. The communication between LV and RV and the narrower pulmonary valve provoke the increases of the RV pressure and hypertrophy of the RV wall that becomes thicker.

TOF typically requires open-heart surgery in the first years of life. The procedure involves increasing the size of the pulmonary valve and pulmonary arteries and repairing the ventricular septal defect. The exact timing of surgery depends on the

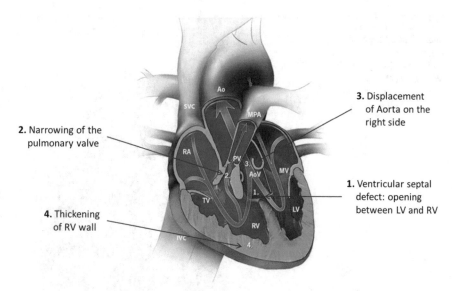

Fig. 13.15 Tetralogy of Fallot (credit: Centers for Disease Control and Prevention, National Center on Birth Defects and Developmental Disabilities, Public domain)

symptoms and size. Normally, surgery is delayed as much as possible in order to act on more grown hearts. When surgery is made early, further surgery may be required to adapt the therapeutic repairs along with the increasing size of the heart.

The big challenge in TOF therapy is, therefore, to be able to anticipate the evolution of the disease, in order to better plan the timing of the various therapeutic activities. The dynamic analysis of intra-cardiac fluid dynamics was recognized to have a role in cardiac morphogenesis as well in cardiac development. Therefore, research is in progress to evaluate fluid dynamics in TOF patients, especially by 3D phase-contrast MRI (4D flow MRI), with more centers under creation in numerous sites. The aim is of providing evaluations of the actual status of the cardiac circulation and, possibly, indications of the probable evolutions that can be precious for optimization of surgical choices and timing of therapy.

The hypoplastic left heart syndrome (HLHS) is a birth defect that occurs when the left side of the heart is underdeveloped; in this pathology, the LV is not formed or it is very small, and the ascending portion of the aorta is typically underdeveloped as well (Fig. 13.16). These babies must undergo surgery immediately at birth because blood is not pumped into the primary circulation.

A baby with hypoplastic left heart syndrome has to undergo to multiple surgeries in a particular order to increase blood flow to the body and bypass the poorly functioning left side of the heart (Love 2017). These surgeries do not cure hypoplastic left heart syndrome, but help to restore heart function aiming to transform the right ventricle in the main pumping chamber to the entire body. Soon after birth, babies undergo the first surgery (Norwood Procedure). This creates a new aorta connected to the right ventricle, such that the right ventricle is used to pump blood both to the pulmonary

Fig. 13.16 Hypoplastic left
heart syndrome (credit:
Centers for Disease Control
and Prevention, CC0, via
Wikimedia Commons)

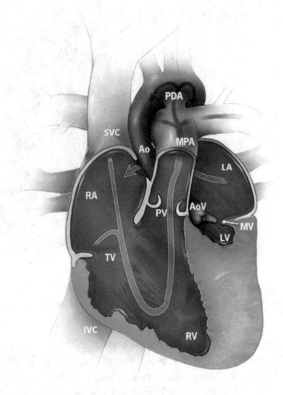

and the systemic circulation. After a few months, an additional surgery is commonly
required (bi-directional Glenn Shunt) that creates a direct connection between the
pulmonary artery and the superior vena cava. This reduces the work of the RV for
the pulmonary circulation and increases it for systemic circulation.

A conclusive procedure (Fontan procedure) is usually performed when the baby
has about 3 years of age. In this final configuration, the inferior and superior venae
cavae are connected directly to the pulmonary artery (total cavo-pulmonary connection, TCPC) and the RV is completely bypassed and not used to pump blood into the
lung. Pulmonary circulation is thus assumed to occur naturally, without the thrust
from the RV, supported by the pressure available in the caval veins. The RV is then
connected directly to the aorta and it is used for systemic circulation. This final circulation, where the RV is transformed in the systemic ventricle, is commonly called
the one with a single right ventricle (SRV). It is evident that the solution is normally
with a reduced cardiac function, the outcome depending on details of flow circulation
in the new configuration. For this reason, it should be recommended that the fluid
dynamics of these patients is monitored.

References

Andersson, M., and T. Ebbers, M. Karlsson. 2019. Characterization and estimation of turbulence-related wall shear stress in patient-specific pulsatile blood flow. *The Journal of Biomechanics* 85: 108–117. https://doi.org/10.1016/j.jbiomech.2019.01.016.

Arvidsson, P.M., and J. Töger, G. Pedrizzetti, E. Heiberg, R. Borgquist, M. Carlsson, H. Arheden. 2018. Hemodynamic forces using 4D flow MRI: an independent biomarker of cardiac function in heart failure with left ventricular dyssynchrony? *The American Journal of Physiology: Heart and Circulatory Physiology* ajpheart.00112.2018. https://doi.org/10.1152/ajpheart.00112.2018.

Barenblatt, G.I. 2003. *Scaling*. Cambridge University Press.

Berne, R.M., and M.N. Levy. 1997. *Cardiovascular physiology*. Mosby-Year Book, Inc., St. Louis.

Bolger, A.F., and E. Heiberg, M. Karlsson, L. Wigström, J. Engvall, A. Sigfridsson, T. Ebbers, J.P.E. Kvitting, C.J., B. Wranne. 2007. Transit of blood flow through the human left ventricle mapped by cardiovascular magnetic resonance. *Journal of Cardiovascular Magnetic Resonance* 9: 741–747. https://doi.org/10.1080/10976640701544530.

Bourantas, C.V., and P.W. Serruys. 2014. Evolution of transcatheter aortic valve replacement. *Circulation Research* 114: 1037–1051. https://doi.org/10.1161/CIRCRESAHA.114.302292.

Carlhäll, C.J., and A. Bolger. 2010. Passing strange flow in the failing ventricle. *Circulation: Heart Failure* 3: 326–331. https://doi.org/10.1161/CIRCHEARTFAILURE.109.911867.

Caro, C.G., and D.J. Doorly, M. Tarnawski, K.T. Scott, Q. Long, C.L. Dumoulin. 1996. Non-planar curvature and branching of arteries and non-planar-type flow. *Proceedings: Mathematical, Physical and Engineering Sciences* 452: 185–197.

Caro, C.G., and J.M. Fitzgerald, R.C. Schroter. 1969. Arterial wall shear and distribution of early atheroma in man. *Nature* 223: 1159–1161. https://doi.org/10.1038/2231159a0.

Carpentier, A., and D.H. Adams, F. Filsoufi. 2010. *Carpentier's reconstructive valve surgery - from valve analysis to valve reconstruction*, 1st ed. Elsevier.

Chen, P.Y., and L. Qin, G. Li, Z. Wang, J.E. Dahlman, J. Malagon-Lopez, S. Gujja, N.A. Cilfone, K.J. Kauffman, L. Sun, H. Sun, X. Zhang, B. Aryal, A. Canfran-Duque, R. Liu, P. Kusters, A. Sehgal, Y. Jiao, D.G. Anderson, J. Gulcher, C. Fernandez-Hernando, E. Lutgens, M.A. Schwartz, J.S. Pober, T.W. Chittenden, G. Tellides, M. Simons. 2019. Endothelial TGF-β signalling drives vascular inflammation and atherosclerosis. *Nature Metabolism* 1: 912–926. https://doi.org/10.1038/s42255-019-0102-3.

Chung, B., and J.R. Cebral. 2015. CFD for evaluation and treatment planning of aneurysms: review of proposed clinical uses and their challenges. *Annual Review of Biomedical Engineering* 43: 122–138. https://doi.org/10.1007/s10439-014-1093-6.

© The Editor(s) (if applicable) and The Author(s), under exclusive license to Springer Nature Switzerland AG 2022
G. Pedrizzetti, *Fluid Mechanics for Cardiovascular Engineering*,
https://doi.org/10.1007/978-3-030-85943-5

Collia D., and L. Zovatto, G. Tonti, G. Pedrizzetti. 2021. Comparative Analysis of Blood Flow in the Right Ventricle of the Human Heart. Frontiers in Bioengineering and Biotechnology-Biomechanics 2021; 9:667408. DOI:https://doi.org/10.3389/fbioe.2021.667408

Courtois, M., and S.J. Kovács, P.A. Ludbrook. 1988. Transmitral pressure-flow velocity relation. Importance of regional pressure gradients in the left ventricle during diastole. *Circulation* 78: 661–71. https://doi.org/10.1161/01.cir.78.3.661.

Cunningham, K.S., and A.I. Gotlieb. 2005. The role of shear stress in the pathogenesis of atherosclerosis. Lab. *Investig.* 85: 9–23. https://doi.org/10.1038/labinvest.3700215.

Davidson, P.A. 2004. *Turbulence: An introduction for scientists and engineers.* Oxford Uuniversity Press, New York.

Domenichini, F. 2011. Three-dimensional impulsive vortex formation from slender orifices. *Journal of Fluid Mechanics* 666: 506–520. https://doi.org/10.1017/S0022112010004994.

Eriksson, J., and J. Zajac, U. Alehagen, A.F. Bolger, T. Ebbers, C.-J. Carlhäll. 2017. Left ventricular hemodynamic forces as a marker of mechanical dyssynchrony in heart failure patients with left bundle branch block. *Scientific Reports* 7: 2971. https://doi.org/10.1038/s41598-017-03089-x.

Faludi, R., and M. Szulik, J. D'hooge, P. Herijgers, F. Rademakers, G. Pedrizzetti, J.-U.J.-U. Voigt. 2010. Left ventricular flow patterns in healthy subjects and patients with prosthetic mitral valves: An in vivo study using echocardiographic particle image velocimetry. *The Journal of Thoracic and Cardiovascular Surgery* 139: 1501–1510. https://doi.org/10.1016/j.jtcvs.2009.07.060.

Feigenbaum, M.J. 1978. Quantitative universality for a class of nonlinear transformations. *Journal of Statistical Physics* 19: 25–52. https://doi.org/10.1007/BF01020332.

Firstenberg, M.S., and P.M. Vandervoort, N.L. Greenberg, N.G. Smedira, P.M. McCarthy, M.J. Garcia, J.D. Thomas. 2000. Noninvasive estimation of transmitral pressure drop across the normal mitral valve in humans: Importance of convective and inertial forces during left ventricular filling. *Journal of the American College of Cardiology* 36: 1942–1949. https://doi.org/10.1016/S0735-1097(00)00963-3.

Fredriksson, A.G., and J. Zajac, J. Eriksson, P. Dyverfeldt, A.F. Bolger, T. Ebbers, C.J. Carlhäll. 2011. 4-D blood flow in the human right ventricle. *The American Journal of Physiology-Heart and Circulatory Physiology* 301: 2344–2350. https://doi.org/10.1152/ajpheart.00622.2011.

Frisch, U. 1995. *Turbulence. The Legacy of A. N. Kolmogorov.* Cambridge University Press. https://doi.org/10.1017/CBO9781139170666.

Fung, Y.C. 1997. *Biomechanics: Circulation*, 2nd ed. Springer, New York.

Gharib, M., and E. Rambod, A. Kheradvar, D.J. Sahn, J.O. Dabiri. 2006. Optimal vortex formation as an index of cardiac health. *Proceedings of the National Academy of Sciences of the United States of America* 103: 6305–6308. https://doi.org/10.1073/pnas.0600520103.

Guerra, M., and C. Brás-Silva, M.J. Amorim, C. Moura, P. Bastos, A.F. Leite-Moreira. 2013. Intraventricular pressure gradients in heart failure. *Physiological Research* 62: 479–487.

Kheradvar, A., and G. Pedrizzetti.2012. Vortex formation in the cardiovascular system, Vortex Formation in the Cardiovascular System. https://doi.org/10.1007/978-1-4471-2288-3.

Kilner, P.J., and G.Z. Yang, A.J. Wilkes, R.H. Mohiaddin, D.N. Firmin, M.H. Yacoub. 2000. Asymmetric redirection of flow through the heart. *Nature* 404: 759–61. https://doi.org/10.1038/35008075.

Kundu, P.K., and I.M. Cohen, D.R. Dowling. 2012. *Fluid mechanics*, 5th ed. Academic Press.

Lee, S.W., and L. Antiga, D.A. Steinman. 2009. Correlations among indicators of disturbed flow at the normal carotid bifurcation. *The Journal of Biomechanical Engineering* 131: 1–7. https://doi.org/10.1115/1.3127252.

Liu, X., and A. Sun, Y. Fan, X. Deng. 2015. Physiological significance of helical flow in the arterial system and its potential clinical applications. *Annual Review of Biomedical Engineering* 43: 3–15. https://doi.org/10.1007/s10439-014-1097-2.

Lloyd-Jones, D.M., and M.G. Larson, E.P. Leip, A. Beiser, R.B. D'Agostino, W.B. Kannel, J.M. Murabito, R.S. Vasan, E.J. Benjamin, D. Levy. 2002. Lifetime risk for developing congestive heart failure: The Framingham Heart Study. *Circulation* 106: 3068–3072. https://doi.org/10.1161/01.CIR.0000039105.49749.6F.

Love, B.A. 2017. Transcatheter Superior Cavopulmonary anastomosis: Interesting technique, limited applicability. *Journal of the American College of Cardiology* 70: 753–755. https://doi.org/10.1016/j.jacc.2017.06.048.

Mangual, J.O., and F. Domenichini, G. Pedrizzetti. 2012. Describing the highly three dimensional right ventricle flow. *Annual Review of Biomedical Engineering* 40: 1790–1801. https://doi.org/10.1007/s10439-012-0540-5.

May, R.M. 1976. Simple mathematical models with very complicated dynamics. *Nature* 261: 459–467. https://doi.org/10.1038/261459a0.

Messner, A.M., and G.Q. Taylor. 1980. Algorithm 550: Solid polyhedron measures [Z]. *ACM Transactions on Mathematical Software* 6: 121–130. https://doi.org/10.1145/355873.355885.

Monin, A.S., and A.M. Yaglom. 1971. *Statistical fluid mechanics*, Volume 1. MIT Press.

Morbiducci, U., and R. Ponzini, G. Cadioli, M. Cadioli, A. Esposito, F.M. Montevecchi, A. Redaelli. 2011. Mechanistic insight into the physiological relevance of helical blood flow in the human aorta: An in vivo study. *Biomechanics and Modeling in Mechanobiology* 10: 339–355. https://doi.org/10.1007/s10237-010-0238-2.

Panton, R.L. 2013. *Incompressible flow*, 4th ed. John Wiley & Sons, Inc., Hoboken, NJ, USA. https://doi.org/10.1002/9781118713075.

Park, K.-H., and J.-W. Son, W.-J. Park, S.-H. Lee, U. Kim, J.-S. Park, D.-G. Shin, Y.-J. Kim, J.-H. Choi, H. Houle, M.A. Vannan, G.-R. Hong. 2013. Characterization of the left atrial vortex flow by two-dimensional Transesophageal contrast echocardiography using particle image velocimetry. *Ultrasound in Medicine and Biology* 39: 62–71. https://doi.org/10.1016/j.ultrasmedbio.2012.08.013.

Pasipoularides, A. 2015. Mechanotransduction mechanisms for intraventricular diastolic vortex forces and myocardial deformations: Part 2. *Journal of Cardiovascular Translational Research* 8: 293–318. https://doi.org/10.1007/s12265-015-9630-8.

Pedley, T.J. 1980. *The fluid dynamics of large blood vessels*. Cambridge University Press.

Pedley, T.J., and B.S. Brook, R.S. Seymour. 1996. Blood pressure and flow rate in the giraffe jugular vein. *Philosophical Transactions of the Royal Society B: Biological Sciences* 351: 855–866. https://doi.org/10.1098/rstb.1996.0080.

Pedrizzetti, G. 2019. On the computation of hemodynamic forces in the heart chambers. *The Journal of Biomechanics* 109323. https://doi.org/10.1016/j.jbiomech.2019.109323.

Pedrizzetti, G., and F. Domenichini. 2015. Left ventricular fluid mechanics: The long way from theoretical models to clinical applications. *Annual Review of Biomedical Engineering* 43: 26–40. https://doi.org/10.1007/s10439-014-1101-x.

Pedrizzetti, G., and F. Domenichini. 2005. Nature optimizes the swirling flow in the human left ventricle. *Physical Review Letters* 95: 1–4. https://doi.org/10.1103/PhysRevLett.95.108101.

Pedrizzetti, G., and G. La Canna, O. Alfieri, G. Tonti. 2014. The vortex—an early predictor of cardiovascular outcome? *Nature Reviews Cardiology* 11: 545–553. https://doi.org/10.1038/nrcardio.2014.75.

Pedrizzetti, G., and A.R. Martiniello, V. Bianchi, A. D'Onofrio, P. Caso, G. Tonti. 2016. Changes in electrical activation modify the orientation of left ventricular flow momentum: Novel observations using echocardiographic particle image velocimetry. *The European Heart Journal—Cardiovascular Imaging* 17: 203–209. https://doi.org/10.1093/ehjci/jev137.

Pedrizzetti, G., and P.P.P. Sengupta. 2015. Vortex imaging: new information gain from tracking cardiac energy loss. *The European Heart Journal – Cardiovascular Imaging* 16: 10–11. https://doi.org/10.1093/ehjci/jev070.

Pijls, N.H.J., and B. de Bruyne, K. Peels, P.H. van der Voort, H.J.R.M. Bonnier, J. Bartunek, J.J. Koolen. 1996. Measurement of fractional flow reserve to assess the functional severity of coronary-artery stenoses. *The New England Journal of Medicine* 334: 1703–1708. https://doi.org/10.1056/NEJM199606273342604.

Qian, J., and Z. Gao, C. Hou, Z. Jin. 2019. A comprehensive review of cavitation in valves: mechanical heart valves and control valves. *Bio-Design and Manufacturing* 2: 119–136. https://doi.org/10.1007/s42242-019-00040-z.

Redaelli, A., and G. Guadagni, R. Fumero, F. Maisano, O. Alfieri. 2001. A computational study of the hemodynamics after "edge-to-edge" mitral valve repair. *The Journal of Biomechanical Engineering* 123: 565–570. https://doi.org/10.1115/1.1408938.

Regueiro, A., and, J.F. Granada, F. Dagenais, J. Rodés-Cabau. 2017. Transcatheter mitral valve replacement: Insights from early clinical experience and future challenges. *Journal of the American College of Cardiology* 69: 2175–2192. https://doi.org/10.1016/j.jacc.2017.02.045.

Reynolds, O. 1894. II. On the dynamical theory of incompressible viscous fluids and the determination of the criterion. *Proceedings of the Royal Society of London* 56: 40–45. https://doi.org/10.1098/rspl.1894.0075.

Reynolds, O. 1883. III. An experimental investigation of the circumstances which determine whether the motion of water shall be direct or sinuous, and of the law of resistance in parallel channels. *Proceedings of the Royal Society of London* 35: 84–99. https://doi.org/10.1098/rspl.1883.0018.

Reynolds, W.C., and A.K.M.F. Hussain. 1972. The mechanics of an organized wave in turbulent shear flow. Part 3. Theoretical models and comparisons with experiments. *Journal of Fluid Mechanics* 54: 263–288. https://doi.org/10.1017/S0022112072000679.

Riley, K.F., and M.P. Hobson, S.J. Bence. 2006. *Mathematical methods for physics and engineering*, Third. ed. Cambridge University Press.

Riley, N. 2001. Steady streaming. *Annual Review of Fluid Mechanics* 33: 43–65. https://doi.org/10.1146/annurev.fluid.33.1.43.

Rubenstein, D.A., and W. Yin, M.D. Frame. 2015. *Biofluid Mechanics: An introduction to fluid mechanics, macrocirculation, microcirculation*, 2nd ed. Academic Press.

Saffman, P.G. 1992. Vortex *Dynamics*. Cambridge University Press.

Sagaut, P. 2006. *Large eddy simulation for incompressible flows, scientific computation*. Springer, Berlin/Heidelberg. https://doi.org/10.1007/b137536.

Schlichting, H. 1979. *Boundary-layer theory*, VII editio. ed. McGraw-Hill.

Seo, J.H., and T. Abd, R.T. George, R. Mittal. 2016. a coupled chemo-fluidic computational model for Thrombogenesis in infarcted left ventricles. *The American Journal of Physiology-Heart and Circulatory Physiology* 310: H1567–H1582. https://doi.org/10.1152/ajpheart.00855.2015.

Tanigaki, T., and H. Emori, Y. Kawase, T. Kubo, H. Omori, Y. Shiono, Y. Sobue, K. Shimamura, T. Hirata, Y. Matsuo, H. Ota, H. Kitabata, M. Okubo, Y. Ino, H. Matsuo, T. Akasaka. 2019. QFR Versus FFR derived from computed tomography for functional assessment of coronary artery stenosis. *JACC Cardiovasc. Interv.* 12: 2050–2059. https://doi.org/10.1016/j.jcin.2019.06.043.

Taylor, C.A., and T.A. Fonte, J.K. Min. 2013. Computational fluid dynamics applied to cardiac computed tomography for noninvasive quantification of fractional flow reserve. *Journal of the American College of Cardiology* 61: 2233–2241. https://doi.org/10.1016/j.jacc.2012.11.083.

Tonti, G., and G. Pedrizzetti, P. Trambaiolo, A. Salustri, 2001. Space and time dependency of inertial and convective contribution to the transmitral pressure drop during ventricular filling. *Journal of the American College of Cardiology* 38: 290–291. https://doi.org/10.1016/S0735-1097(01)01355-9.

Ziegler, M., and M. Welander, J. v, M. Lindenberger, N. Bjarnegård, M. Karlsson, T. Ebbers, T. Länne, P. Dyverfeldt. 2019. Visualizing and quantifying flow stasis in abdominal aortic aneurysms in men using 4D flow MRI. *Magnetic Resonance Imaging* 57: 103–110. https://doi.org/10.1016/j.mri.2018.11.003.

Printed in the United States
by Baker & Taylor Publisher Services